Praise for
The Sonic Boom

"Thanks to such entertaining asides and Beckerman's enthusiastic conviction that sound matters more than anything, readers of *The Sonic Boom* might suddenly find they're hearing things they had never noticed before." — *Washington Independent Review of Books*

"The only force more powerful than taste or smell when trying to create an emotional response? Music. The right sound at the right time can slam you right back to childhood, create dread, inspire empathy, turn you on a dime to an emotional jellyfish. The right song at the right time can change — and has changed — the world, leading directly to social change and even revolutions. Music defines us. Joel Beckerman knows. Let him tell you all about it."
    — Anthony Bourdain, chef, best-selling author of *Medium Raw,*
    No Reservations, and *Kitchen Confidential*

"*The Sonic Boom* will alter how you hear the world." — *Time*

"Pick this book up. Put it close to your ear. Riffle through the pages. Hear that? Those are ideas — Beckerman's ideas — flying by a mile a minute. Reading *The Sonic Boom* is like seeing colors for the first time!" — Fred Graver, writer/producer, TV creative lead at Twitter

"Beckerman's anecdotes, including how Apple computers came to have their particular start-up noise, are engaging."
    — *Los Angeles Times*

"Offers an intriguing examination of the manipulative and inspirational power of sound in our everyday lives. It made me listen to my ears!"
— George S. Clinton, film composer and chair of film scoring at Berklee College of Music

"We are all susceptible to the power and manipulation of sound, and Beckerman has beautifully described this mysterious process."
— Gordon Elliott, executive producer of ABC's *The Chew*

"This remarkable book describes and clarifies the exciting and complex world of sound." — Ellis Douek, author of *Overcoming Deafness*

"Beckerman offers insights into a potent marketing opportunity. His work helps brands realize the powerful potential of sound."
— Ruth Gaviria, executive vice president of corporate marketing, Univision Communications, Inc.

"Illuminates the surprising opportunities that are revealed when we utilize sound thoughtfully and strategically, not just tactically."
— Andrea Sullivan, chief marketing officer, North America, Interbrand

"Informative . . . This book is directly aimed at corporations, fundraisers, and party planners, but shoppers, donors, and partygoers should also read it to learn about how their decisions are being affected by the soundscapes in which they are immersed."
— *Publishers Weekly*

"A fast and provocative read that will appeal to a wide, general audience, especially those interested in the psychology of marketing." — *Library Journal*

THE SONIC BOOM

How Sound Transforms the Way
We Think, Feel, and Buy

JOEL BECKERMAN

WITH TYLER GRAY

Mariner Books
Houghton Mifflin Harcourt
BOSTON   NEW YORK

First Mariner Books edition 2015

Copyright © 2014 by Man Made Music, Inc.

For information about permission to reproduce selections from this book, write to
trade.permissions@hmhco.com or to Permissions, Houghton Mifflin Harcourt
Publishing Company, 3 Park Avenue, 19th Floor, New York, New York 10016.

www.hmhco.com

Library of Congress Cataloging-in-Publication Data
Beckerman, Joel.
The sonic boom : how sound transforms the way we think, feel, and buy /
Joel Beckerman ; contributions by Tyler Gray.
pages cm
ISBN 978-0-544-19174-7 (hardback)   ISBN 978-0-544-57016-0 (pbk.)
1. Music — Psychological aspects. 2. Marketing — Psychological aspects
3. Music in advertising. 4. Sound — Psychological aspects. I. Title.
ML3830.B33 2014
781.2'3 — dc23
2014016521

Book design by Brian Moore

Printed in the United States of America
DOC 10 9 8 7 6 5 4 3 2 1

"Mister Softee" lyrics are reprinted by permission of Mister Softee Inc.

Dedicated to the remarkable souls who inspire me to create and truly listen, and to Tracy, who loves me even though I conduct in my sleep

Contents

Introduction

THE WAY WE HEAR NOW

Imagine you've just stepped into a very popular modern American casual-dining restaurant. I'll tell you which one later in the book, though you'll probably be able to figure it out for yourself. It's the kind where the aroma of onions smacks you in the face the second you pull open the doors, where Western-themed memorabilia adorns the walls. And when you arrive at your seat, you're barraged with brightly colored menu pitches for frozen specialty drinks and gooey desserts.

Here comes the important part.

Just a few moments after you start perusing the menu, you hear a hiss from behind double doors — the kitchen. When the doors burst open, the hiss becomes a distinct sizzle. It cuts through the overhead music, the white noise of conversation and laughter, the tinkling glasses, and the rattling ice in the bartender's cocktail shaker. It's startling. And it makes you turn your head. The standout sound careens around tables like an accident in progress. There's an anxiety to it. By now, you're completely focused on tracking where the noise is coming from. You spot steam and smoke. It makes you notice anew the fried-onion smell pervading the restaurant. You don't know what spices are used in the dish or what meat, if any, is involved, but your mouth waters, and somehow you have a sense of how it tastes. You're curious. And curiosity is only a short leap away from craving. Forget the menu — you're already devouring this dish in your mind.

Sizzling fajitas are a novel but powerful everyday example of the hidden power of sound. The auditory input comes first in every sizzling-fajita experience. That sound summons feelings of excitement, joy, and anticipation and rallies and heightens other senses in a chain reaction, pulling in sights, smells, and eventually taste. It makes you instantly feel a story — a fresh, hot, cowboy-style southwestern dish that's prepared to order just for you. The sizzling skillet can influence your choice if you hear it while you're still perusing the menu (or it can make you regret ordering something else if you hear it too late). The experience of having that loud dish delivered to your table sticks with you long after the taste and aroma fade.

The sound of sizzling fajitas is also a powerful tool for business. While I was writing this book, I asked people what restaurant chain came to mind when I mentioned *sizzling fajitas,* and almost everyone named the same place. You're probably thinking of the same place too. (If you're still not certain what it is, you can take a peek at the beginning of chapter 2 for the answer.)

The response to sound is central to the human psyche. It's essential to our humanity and day-to-day experiences. It frames every moment of every day. It shapes our moods, our preferences, and our personal and collective histories, and it triggers memories and powerful emotional reactions and connections. And it does so invisibly.

Imagine your mother singing you a lullaby — do you feel instantly comforted and relaxed, and maybe even a little sleepy?

How about the song that was playing during the first dance at your wedding or at your prom — does it still have the power to bring back the thrill of the moment?

What does imagining the roar of a stadium crowd instantly do to your heart rate?

What do you feel when you hear the theme to *Mission Impossi-*

ble? *Star Wars*? *SpongeBob Square Pants*? When you hear "Thaaaaah Siiiiiiimpsoooons"?

What happens when you hear the tinkly music of the ice cream truck? Does it make you think of the heat of summer? The chill of a creamy sweet treat or faux-fruit refreshment on your tongue? Or maybe it takes you back to a childhood in the suburbs, and you feel a twinge of anxiety from hearing that faint music, which meant begging your mom or dad for change and sprinting out the door before the truck rolled by. Sound initiates all those feelings and memories of sights, temperatures, and tastes.

Think about what happens when you walk into a Starbucks. Even before the coffee aroma grabs you, it's likely that you hear the hiss of the milk steamer or the bang of the espresso portafilter being dumped for the next batch. There's also the distinctive music that Starbucks plays and sells in their stores. Your brain is fitting all those sounds into patterns you know — sonic memories and expectations — and combining them with the sights and smells to create a multisensory reaction. But the sound does the work without your realizing it. If you had to consciously consider all of this stuff, you'd be exhausted by the time you hit your desk at nine o'clock, whether or not you'd downed a venti latte.

Music in particular helps an experience become a memory and later helps you recall those memories with just a few of the right notes. Ever wonder why you struggled to learn the names of all the U.S. presidents but can sing the entire process of how a bill becomes a law? ("I'm just a bill . . .")

Why is it that you probably can't name the capitals of all fifty states but you can recite a significant portion of the Oscar Mayer wiener song?

Try to recite the alphabet without hearing the music — or at least the *el-em-en-oh-pee* cadence — of the ABC song.

In my musical, professional, and personal life, I often ask people questions like these to demonstrate all the instances where sound and music drives our reactions. This ear-opening exercise affects new converts — whether they are friends, family, acquaintances, collaborators, or clients — the same way almost every time. Once they know the basic ideas, they start to hear the world in a brand-new way. Something powerfully unconscious becomes powerfully conscious. They're aware of a world of sounds around them that they never paid attention to before. They hear things they never heard. They make connections. And when they come back to tell me about it, they are always wide-eyed and smiling. "I hear this *everywhere* now!" They suddenly realize that the reason they always feel so irritable in the grocery store is the horrible music being piped through the speakers, and they understand why they feel that sharp pang of excitement when the ping of the phone announces a new message.

Why would all these smart people not be aware of the powerful role of sound already? Because it's so pervasive, they scarcely notice it. Sound is present every moment of our lives, affecting our moods, our reactions, our thoughts, and our choices on a largely subconscious level.

I've been a musician since I was seven. After a gospel piano player came to visit my grade school, I begged my parents for piano lessons. I desperately wanted to learn how to make that kind of sound. But it wasn't until I was eleven that I recognized sound and music's power to shape, define, and transform an emotional experience. I had to show my little brother that I could get through *The Exorcist* without wetting myself, so I used sound to feel brave. It was way too late at night. The babysitter was conked out on the couch. I dialed up the classic horror flick on cable. During graphic scenes, I didn't cover my eyes. Instead, I turned down the sound and replaced it with the tones of our old portable organ, the kind that had a humming fan that blew across whistle-like plastic reeds to make the notes. I came up with my own impromptu live score for the pictures on the screen. As the cheesy organ chords filled the room, Linda Blair's spinning head seemed silly. My brother caught on and reached through a hole in the nearby 1970s-era beanbag chair on which my sister had long since fallen asleep. He came up with a fistful of plastic beans, stuffed them into one side of the organ, and let the fan blast them out the other end. As Linda spewed pea-green soup, we rolled in a plastic-bean blizzard, laughing until tears streamed down our faces.

The babysitter woke up and was not amused.

But that was just the beginning of the fun. By dialing down the TV volume and cranking up the hi-fi, I could make something funny become spooky, morph lighthearted shows into dark dramas, and turn major world events into slapstick. Ever see the YouTube clip (millions have) in which YouTube user Neochosen replaces the music in the trailer for Stanley Kubrick's classic psycho-thriller *The Shining* with Peter Gabriel's "Solsbury Hill"? Suddenly, Jack Nicholson's ax murderer becomes a quirky, doting father in a feel-good rom-com. That was the same concept my siblings and I played with decades ago. Then it was all just fun and games. I didn't realize until much later that when my brain received contradictory input, it always believed the sound and made that the anchor for the experience.

In the nineties, my musical play morphed into work. By day, I toiled at Manhattan's SOJ Studios, in Midtown West, as a producer and engineer on publishing demos for songwriters. Over the course of a couple of years, I recorded probably two hundred songs. I'd have five hours to crank out a full song with production. It was a constant crash course in playing, producing, arranging, and recording. I learned from my heroes: record producers like Berry Gordy, Teo Macero, Dr. Dre, and Quincy Jones; master songwriters like Paul Simon, Otis Redding, Lennon and McCartney, and Johnny Cash; great concert music composers like Ravel, Wagner, Bernstein, and Berg; film composers like Bernard Herrmann, Jerry Goldsmith, and John Williams. In my publishing-demo work at SOJ, I learned through trial and error the real craft of music-making in the studio: how to structure a song so the hook pays off and sticks in a listener's mind, how a great lyric can be completely shattered by one wrong word or the wrong delivery, and, perhaps most important, how not to get too enamored of production — to get out of the way of the song and let its meaning and emotion shine through. I learned, essentially, to trust the song.

In the evenings, I'd shoot crosstown to Midtown East, where I was the night manager of another studio, HSR. Plenty of TV-spot soundtracks and records were made there, but what caught my attention were the amazing radio spots recorded by one of my mentors, Richie Becker. I listened closely as sonic artists like Richie produced ads for clients such as Mercedes-Benz, United Airlines, National Car Rental, Burger King, and others. They created these completely immersive mental pictures, using sound to make listeners feel things, like the wind on a racetrack or the thrill of taking a high-performance car around a mountain curve. (Richie is now the evil genius behind the sound of Fox Sports.) Highly skilled voice actors, perfectly delivered scripts, and just as you thought the piece might be ending: *vroom!* The rev of the Mercedes engine came back for a sonic surprise. Sound made you feel and want a car and an experience that you couldn't even see on the radio. For me, it opened up a whole new appreciation for the power of sound—not only in the music I recorded by day but in the sounds I heard used to create theater of the mind in the evening. For the first time I saw how I could use a whole range of sound and music to tell emotional stories.

Today, I'm fortunate enough to be a busy composer and producer, but I'm still a student of music and sound, endlessly fascinated and astonished by its power to tell stories and move people. And I'm no less in awe of those masters who move us and transform our experiences with sound. Sound or music can make powerful emotional connections in an instant; it can bring huge groups of people to tears or fits of laughter, even when those people are miles apart from one another. Sound can instantly change people's moods or perceptions and bring distinct images to their minds. If a picture is worth a thousand words, then the right sound at the right moment is worth a thousand pictures.

Of course, there's no way to describe all of this in the space of, say, a cocktail conversation when someone asks me what I do for a living.

The short answer is that I'm a composer and producer for television and a strategic sonic-branding consultant, though that second part seems to provoke more questions than it answers. Honestly, I'm just a tireless student of the power of music and sound and their effects on our lives. I wanted to help people use the power of sound to score stories, not just on the screen but everywhere, so in 1998 I created a company, Man Made Music, to support that work.

The company attracted uniquely talented arrangers, writers, producers, musicians, club DJs, and sound designers, as well as music supervisors and recording engineers and, later, brand strategists, UX (user experience) designers, programmers, musicologists, and brand-partnership experts. Each of them has brought exciting new dimensions to telling great stories with sound.

With them, I've practiced the art and business of using sounds to turn toys and shoes and gadgets and games into meaningful talismans and lifestyle statements — after all, brands try to tell stories like Hollywood does, and Hollywood looks to brands to bolster business. Strategically deployed, sound can become a potent tool for brands, those groups and individuals desperately seeking to create genuine experiences and engage with their audiences. But ultimately, this book is about something much bigger than me, my company, or commerce; it's about how anyone can harness the power of sound to make his or her life and the lives of others better.

This book explains how to approach and explore sound in a more strategic, holistic way. It will introduce you to the art of curating and placing sound to make the biggest impact and heighten your listeners' emotional connection with the story you're trying to tell. It will also show you how to pull out sound and use silence to accomplish the same storytelling goals. This approach is built on the best practices I've learned from a lifetime of experiences, from the thirty years I've spent working as a composer and producer for television and other media, and from the brilliant musical and creative minds I've had the honor of collaborating with. I've had the opportunity to help some very big brands fine-tune their messages with sound, designing AT&T's sonic logo (those four little notes you hear at the end of every AT&T ad), creating Univision's theme song and turning it into an anthem for a movement, and scoring the cinematic drama of the Super Bowl. I'll share some of what I learned from those experiences later in the book.

Any size business — not just giant corporations — can harness the power of sound to make these kinds of connections. I'll share examples of how effective sound can help you alter your mood, make an impression at a job interview, set the tone for a fundraising event, or help bolster an employee's or client's understanding of your business.

This is not just a big idea for businesses or marketers. *The Sonic Boom* is about harnessing the power of sound for your life, then scaling it to help advance your story, your message, or your goals. I'll show you how to transform an ordinary pitch into a multisensory experience and an emotional connection. Once you are aware of the true power of sound in your life, you'll never hear the world the same way again.

Many fascinating books have focused on the neuroscience and psychology behind how music works in the brain. Plenty of smart authors and researchers look to colorful fMRI scans and controlled experiments to explain our common humanity by way of the gray matter in our heads. Neuroscience and psychology definitely support the strategies you'll read about here. This book covers enough of that stuff to do right by the academics who spend lifetimes decoding it, but I'll also make the case that science tells only a very small part of the story.

Brain science raises more questions than it answers about sound and emotion. Cognitive neuroscientists (and cognitive psychologists and sociologists) will tell you that they're just starting to understand all the ways sound works, and the brain science grapples only with the tiny part of the issue that can be measured. The rest is rooted much deeper than the brain. It's in something more like the soul or human nature. I don't claim to have cracked this code. And you won't find any suggestion in these pages that one can use sound in the real world for mind control. If that's your idea of how sonic branding works, you need this book more than anyone. (Also, if you're spending your money on "neuromarketing," I have a bridge — and a verse and a chorus — I'd like to sell you.)

Your sonic strategy — your intentional, informed plan for using sound — is more likely to involve a set of best practices and the right inspiration rather than an algorithm or a rainbow-colored set of fMRI images. What I'm pushing isn't brain science. It's sonic humanism. I believe philosophers, artists, and creators who think in the language of sound have even more to tell us than the latest science does about the rapturous impact of sound and music on our emotions and memories. But that doesn't mean it's beyond the average person's understanding or ability to apply it. Quite the contrary. This is information you can *use,* even if you can't play an instrument or carry a tune. As human beings, we are wired for this stuff.

Sound can convey troves of information in an instant. It can trigger emotion, then reaction, interaction, and transaction, but only when used properly and with a great deal of integrity. As thrilled as I am when music and sound are applied well, I'm equally appalled when music and sound are used incorrectly — a song tacked onto a commercial or movie to make something cool by association; cacophony that attracts notice through tricks and sheer volume but offers no emotional connection. We live in a blinding, overwhelming world of sensory warfare. Visual, tactile, and auditory stimuli are constantly battling for our attention. When it comes to sound, many are under the impression that more is better, but this book shows you how much more powerful many stories and experiences can be when you pull the sound out. Music and sound must be part of a thoughtful storytelling or communication strategy, not just a tactic. If the sound is doing nothing but driving the visual presentation or adding energy or attempting to make something seem cooler than it actually is, then all it is is noise.

Audiences can sense a lack of integrity a mile away. You might be able to fool people once or twice, but if you constantly lie to them with music and sound, they'll stop trusting you. Think about canned laugh tracks — when one is played after a corny joke no real humans would find funny, the obvious lack of authenticity will make an already bad sitcom seem worse. Music and sound create either a positive connection or a negative disconnection. They're never neutral. And when music doesn't ring true or match the message, it's feels like an offer of friendship from a pathological liar.

In 2003, for example, Nissan put the song "Gravity Rides Everything," by seminal indie rock band Modest Mouse, in a commercial for its minivans (along with images of a surfing mom stowing her board in the back). Fans of the band were not fooled into believing the minivan was cool. Comments on message boards included statements like *I'm not drunk yet, did I just hear that?* and *The four horsemen have saddled up.* In 2010, Honda also linked rock music and a minivan, but it tapped into its Gen-X audience's sense of irony with over-the-top heavy-metal music and images of its vans alongside leather-clad hair-metal bands, flames, and fog. Sound led a spoof of the idea that the minivan is cool, and in doing so, it helped Honda convey the true story of its product: it's for people who once rocked, even if their vehicles are now considerably less . . . rocking.

This book will show you how to spot when bad sound is affecting you and teach you what you can do about it. Then you'll learn how to avoid telling the wrong sonic story.

But first, we'll cover the basics of how sound works, both in branding and in everyday scenarios. You'll learn how to open your ears to the sounds around you and how to pay attention to how they affect you. Some of the most powerful experiences are stronger when they start with sound — no other sensory input is as efficient when it comes to triggering these complex reactions. You make decisions based on subconscious sonic information when choosing what to eat, where to sit, what to wear, how to feel, and more. When the right sound or music is deployed at the right moment to communicate this information and emotion and help you recall memories and forge new experiences, you get what I call a boom moment. I'll help you spot them and show you how to create them.

For businesses seeking these boom moments, sound and music need to be meaningful and authentically tied to the story of their brands. Marketers and creative-decision-makers shouldn't choose music just because they like it. Effective use of sound means building a strategy to pick and create music that helps people understand the brand's place in their world. With the right foundation, sound and music can help transform a business or a message by communicating a clear emotional story and helping people *feel* the brand.

This same principle holds true outside the business world. Sound and music can become the emotional engine that galvanizes a movement. They can rally people around a shared set of values. They can even bridge distances between disparate ethnic and national groups. On a more personal level, they can change the way you interact with your environment and influence the way you are perceived by others.

I have been fortunate to have brilliant colleagues, friends, and teachers show me how all this is done. Innumerable sonic masters' work has inspired and dramatically affected me. It's not possible to thank them all, though I mention some in the following chapters and some in the acknowledgments. They're spirited educators, legendary composers, musicians, singers, songwriters, creative recording engineers, scriptwriters, filmmakers, and editors. I'm sharing what I've learned from marketers, music mavens, and sonic enthusiasts, but also what I've gleaned from artists, physicists, and church-bell ringers, not to mention infants, animators, theme-park-experience experts, and rabid music fans.

There are lots of books for audiophiles and musicians out there; this book is for anyone with ears. You don't need to be able to read music or name all the instruments in an orchestra to harness the power of music and sound. You don't need to know the name of every Beatles song, and if you can't sing your way out of a paper bag, that's fine. You may not think you have a musical bone in your body, but if you've ever had an emotion or a feeling, you have everything you need to put the tools in this book to use. You're a human being. You're wired for this stuff, even if you've never noticed it before. Even some musicians and music lovers may not be fully aware of just how powerful sound can be or of how it can transform so many experiences. The quickest way to understand the power of sound is by seeing what happens when it's gone.

Night-Vision Goggles for Your Ears

O N SEPTEMBER 26, 2011, Sarah Churman, a twenty-nine-year-old mother of two from Fort Worth, Texas, got into her car with her husband. When she closed the door, she says, the noise of it shutting sounded like a bomb going off. Sarah grabbed her phone and dialed her mother-in-law. During the call, one of her two daughters hopped on the line. "Hi, Mommy, I love you," she said. Sarah started bawling so hard that she couldn't catch her breath.

When her husband, Sloan, pulled off the highway later and Sarah opened her door, she said the sound of traffic was roaring. The horns honking and engines revving were thunderous. The two made their way into an Outback Steakhouse and got a table. When the waitress put their drinks down, Sarah jumped at the startling sound. When she began to eat her salad, the crunch of the lettuce was so powerful she couldn't hear what her husband was saying. The noise in her own head was like a cheap used-car commercial with the sound turned up to eleven. It was almost deafening.

Sarah was born with profound hearing loss. For years, she hadn't been able to hear anything quieter than eighty-five decibels. You could rev a gas-powered chain saw or shoot a .22-caliber rifle right next to her ear, and it would sound to her like a mumble. Then Sarah had a device surgically implanted in one ear. It processed sound vibrations, amplified them using the middle-ear bones, and sent them off into Sarah's brain, a task most people's ears handle naturally with tiny hairs (which Sarah was born without). Even in the most successful cases, recipients of the kind of implant Sarah got don't perceive sound quite the same way as people born without hearing problems do. But it's the next best thing. Her mother-in-law used the money she'd saved for retirement to pay for Sarah's surgery. Eight weeks after it, when Sarah's ear had healed, a technician turned the device on, and Sarah began to hear for the first time in twenty-nine years. Sarah truly appreciates the power of sound, something most of us take for granted.

If her name sounds familiar, you might have seen Sarah's first moments of hearing on YouTube or on Ellen DeGeneres's daytime talk show. Sloan captured the event on video and uploaded it. Then it went viral. By mid-2013, more than twenty million people had watched the clip: "Twenty-nine years old and hearing myself for the first time!" It's a heart-melting scene. The slender, brown-haired woman with a full sleeve of tattoos on her right arm stares doe-eyed at a technician, who slowly dials up Sarah's equipment. There's some beeping that Sarah hears first, and then you can see the full breadth of sound wash over her as the implant comes alive. She'd worn hearing aids most of her life, but she described them as always having a constant hum or white noise in the background and allowing her to hear only the loudest external sounds, and even then, every sound blended together so they all sounded muffled or garbled, like Charlie Brown's teacher. This sound was different: clear, bright, and loud as hell. In the YouTube video, she tries to play it cool when the device starts working. That goes to pot in about one second. The flood of emotion that comes with her new sense is almost too much for her to bear. She's self-conscious, so she covers her mouth, which is the way a proficient lip reader like her tries to hide what she's feeling. When Sarah starts shaking and tears start flowing, though, the emotion is obvious. And it's contagious. You can't help but imagine what's happening in her head.

In her memoir *Powered On,* Sarah shares a perspective you can't get from the video alone:

> I realized I could hear the noises coming from my mouth. Then I realized how I sounded, and I got choked up. Then I laughed, and that sent me into a fit of tears. All these sounds were intensified because I was hearing all of this from inside myself for the first time, and I was completely and utterly overwhelmed like you cannot imagine. I feared my heart was gonna explode, and I just couldn't put into words what was racing through my mind.

"I don't want to hear myself cry," she says in the video. Then she laughs and surprises herself. "My laughter sounds so loud!"

"You'll get used to that over time," the technician says.

But it took a while.

It's impossible for a person born with hearing to know what it's like to suddenly get it after twenty-nine years. Even beginning to understand how much sound is all around you is an eye-opening experience. Ever feel frazzled and then realize the TV is blaring an annoying ad or news or a program you don't actually care about? That moment and the next one, when you turn off the TV, is a very basic example of soundscaping. That's the practice of identifying and controlling the sounds affecting you — the loud ones in the foreground, the slightly lower, less important ones in the midground, and the foundational rumblings or whisperings in the background. Some people call the act of noticing these various levels of sound "active listening." Soundscaping is something more. It involves determining which of those sounds are most useful to you, whether you're trying to find your way, figure out what's going on around you, tell a story, or change your mood. Put to use in all kinds of everyday situations, soundscaping is how you use the identification and curation of sound and music to make your world work better. In a way somewhat similar to how Sarah began to hear new sounds and understand the messages they carried, you can grow your sonic vocabulary, the palette of sounds you recognize and use to tell stories or design environments. You can also use this sonic vocabulary to create your own messages and impressions. To get started, you can do a little exercise that will give you a sense of how much sound you're missing at any given moment: Close your eyes and listen to your surroundings, wherever you are. As you do, think about the word *wind*. Do you suddenly hear air moving against things? Next, think about your own ears ringing. Do you immediately hear it?

Now close your eyes again. This time, keep them closed for a full two minutes and try to identify each tiny sound you hear. In the first few seconds, you'll probably notice the loudest sounds first: the music on the television in the other room, a door closing next to you. That's the foreground. But go deeper. Sit with it for a while. Maybe you hear highway traffic a mile away, or the sound of kids outside playing. That's the background. The midground is everything between the background and the foreground.

The longer you listen and the more you pay attention, the more you'll hear. You're not activating some kind of superpower or bionic hearing. You're not even hearing these sounds for the first time. These vibrations have been picked up by your ears all along. You just may not be used to paying attention to them. People who practice yoga or certain types of meditation might be more familiar with identifying sounds, but they may not have considered how these sounds might be affecting them in every moment or how these sounds might ultimately be curated, in the same way a museum director picks a mix of art for the walls.

You can discover lots of information in those sounds. They can make you aware of someone selling you something or explain why you feel certain ways at certain times in certain places. When you become aware of so many more sounds, you start to understand yourself better, and you might even start to use more of your brain in the process. Even if you've always been able to hear, there's plenty you might be missing

Sarah's brain wasn't trained to bring certain sounds to her attention while filtering out less important ones. In the first few hours after her implant was turned on, she heard just about every sound most of us tune out. Sloan tried to remind her of what the doctors had told her: Her brain was firing in all kinds of new ways, and it would take time for her to adjust. Unlike most of us, she didn't have a rich catalog of sonic memories to draw from. Most of us can quickly decipher meaning in ambiguous statements based on tone. We know when someone's joking or being ironic. Not Sarah — she found sarcasm baffling. She'd also never heard her children speak and didn't know how to pick their voices out of a crowd; on the playground, every cry of "Mommy!" seemed directed at her. She was building almost everything from scratch. Sometimes she'd think her truck engine sounded off kilter, even though it wasn't. She'd hear her girls and dart upstairs thinking there was something wrong with them, but they were just playing or being loud. She was startled when the air conditioning kicked in. "Public toilets are insanely loud and make my heart pound every time I flush one. Loud restaurants or bars wear me out," she says. The experience sometimes proves too intense for people who get cochlear implants, and they choose to have the devices taken out. Pretty quickly, though, Sarah began to revel in it all. "I just have more of an appreciation for such things. My dogs grunt and snort when they breathe (I have bulldogs). Cabinet doors squeak. The microwave is loud while it cooks. The fridge runs and makes noise," Sarah said in an interview for this book. It took some time for her to figure out which sounds to pay attention to and which ones to ignore. She had to decide which ones told her stories, communicated valuable information, or helped her know how to feel, judgments most of us make every second of the day without realizing it.

"I pay more attention and take time to listen to them even though they are typical 'mundane' things to others," Sarah says. The irony is that in learning how to pay attention to certain things, Sarah was already training her brain to tune most things out — otherwise, every day would have been like those first few chaotic hours after she'd had her aural implant dialed up. Now, she says, even candy wrappers seem delightful. "I can now hear my phone no matter where it is. A big one was buttons in a car, when you push them they make noise. Buttons on a lot of things make noise!"

In fact, everything makes sound, even the moving of liquid and gas through your internal organs, the friction of air moving across your inner ear, and the fluid in your joints. Since the beginning of time, the universe has whispered and roared. From the cosmological to the quantum level and beyond, we find energy constantly vibrating. This vibration is the root of all sound. It's innate. It's primal. And almost every animal hears or at least senses it. There are plenty of vertebrates who are born without the ability to see or smell or taste. But nothing with a backbone is naturally born without the ability to detect sound vibrations in one form or another. Humans are wired to hear even before they're born; an infant knows its mother's voice before anything else. Archaeologists report finding music in all prehistoric societies of *Homo sapiens*. "With the exception of those who suffer from a cognitive deficit, all individuals have a capacity to acquire language and are born with an inherent appreciation of music," writes Steven Mithen in his 2007 book *The Singing Neanderthals: The Origins of Music, Language, Mind, and Body*. Long before sound activates the rational mind, it dramatically shapes our reality. Sound helps us intuit friend or foe, danger or delight, joy or despair. It's the most visceral, potent way we reach out and connect with one another and our world.

Sound is at the core of our belief structures. The Bible states, "In the beginning was the Word and the Word was God and the Word was with God." Hindus and Buddhists hold that the sacred syllable *om* is the primordial sound of creation. Sound itself, however, is agnostic. It's the hook and trigger in every able-eared human's most cherished memories and most basic emotions. It makes each of us feel something in an instant.

There's a perfectly rational reason you don't think about all of this: you don't have to. People are constantly having sonic experiences, whether they know it, choose it, or even like it. Sound shapes your mood, your energy, and your consciousness, influencing your life and your decisions. But most of us have no idea of the impact of sound, because it lurks below conscious perception. Sound is like air: It's always around you, but you never think about it unless it's taken away.

That's really how a woman who learned to hear after twenty-nine years figures into this book. Sarah gained more than a sense of hearing. She's in the unique position of being able to describe eloquently what it was like to hear for the first time. You share more in common with Sarah than you might think.

You too can learn to hear the hidden world of sound that's influencing you every day. This awareness is like night-vision goggles for your ears. Once you turn it on, you'll see sound hard at work in all sorts of unexpected ways. Think about the motivating sounds of jingling coins in Las Vegas casinos. Or the sound of slot machines — *ka-ching!* — that we literally translate as "money." (By the way, studies have shown that the music and sounds of slot machines make people overestimate how much they're actually winning by as much as 24 percent.) The driving squawks of the Angry Birds and the satisfying crunch of wood and shattering of glass in the Bad Piggies' castles are a huge part of what makes the massively successful app so addictive. Those opening notes over the title credits are part of what make *Star Wars* more than a movie, and those two ominous bass notes from the *Jaws* theme are enough to make anyone think twice about swimming in the ocean. The *Jaws* theme, in fact, is a great example of a sonic logo (although describing it as one of the greatest motifs in film history would be equally appropriate). Those two repeating notes, building in speed and intensity, instantly call up a rich story and all of the emotions (especially fear) associated with it. It might even bring to mind the visual logo for the film, to which it is analogous. You'll often find the sonic logo of a brand at the ends of its ads (though if that's the only place you're hearing it, the brand is missing an opportunity).

Sonic logos are even more powerful when they're tied to anthems. An anthem is the long-form expression of a nation, a brand, a personal story, a movement, or a cause told in the language of sound. It expresses values in a sort of ownable sonic DNA. That DNA can then be used to make shorter sounds — sonic logos — that instantly and efficiently let listeners recall and understand rich stories.

You might not realize it, but you frequently make choices about the spaces you're in based on sound. You may prefer a seat in a corner of a restaurant because it's quieter, or seek out a loud bar because the noise makes you feel like you're part of an exciting scene. Next time you're in a mall or shopping district, listen to what you hear as you walk by each store. The smart businesses tailor their soundtracks the same way better stores tailor clothes. They've identified the values of their target consumers. You're well aware of how a company like the Gap represents those values in its visuals — the stacks of perfectly folded clothes, the clean, bright presentations, and the iconic logo (when the Gap tried to tinker with its logo in 2010, there was such an uproar that the chain quickly changed it back). The language of the Gap's music has to match the same values you see represented visually. It's on trend but never trendy, upbeat but never aggressive in a way that would offend anyone. Like the logo, the music speaks for the brand; you might not realize how much it matters unless it goes horribly wrong — imagine stepping into the Gap and being assaulted by speed metal or booming rap.

Other stores differentiate their brands with sound too, especially in a mall, where there's a different brand experience every fifty feet. You might hear orchestral music spilling out of Brooks Brothers or some kind of melodic metal or aggressive dubstep coming from Hot Topic. What story do the tunes tell you about each store's customers?

Not just that they wear tan pants or guyliner; what does it say about who they are, what matters to them, what their lifestyles might be like? How do the sounds differentiate one store from another?

Now you're ready for the final challenge: How do these sounds make you feel? Do they make you feel anything? That's really at the heart of what I call a boom moment. It's the moment when sound pulls emotional triggers, the instant when sound sets off reactions not just in the parts of your brain that handle auditory stimuli but in the sections associated with memories, fear, joy, and even visual perception and physical sensation or movement. But you don't have to get the neuroscience or psychology or be wired up to an fMRI to understand a boom moment. You'll feel it when a sound makes you remember a certain time of your life or just makes you feel happy, sad, or scared. Or it might make you relate to someone or some-thing—to envision yourself as a character in the story being told. It makes an otherwise ambiguous or meaningless scene, space, or object take on significance or value to you. You probably don't yet realize that it's sound making most of this happen, but the next time you find yourself moved by a gadget or an ad or a scene in a movie, think about whether it would be the same without sound. Try watch-ing *Jaws, Ocean's Eleven,* or any James Bond film with the sound off. It's just not the same. Once you've learned to identify the way these boom moments work, you can start to see sound as a tool, just as Hollywood has for decades.

Boom moments occur when businesses and people use sounds to rope in a lot of other senses, spark memories, tell rich stories with incredible efficiency, and, most important, elicit *feelings.* Often these moments are built on sounds that you don't realize you're hearing. Once you recognize this, you'll be astonished by how often sounds guide your behavior or simplify everyday tasks. Boom moments help everything make more sense. They represent massive sonic opportu-nities for businesses. Many of them are missed, and they're missed by very smart people.

Boom moments are already part of your everyday routine. In New

York City, for example, as you walk down the stairs to the subway, you can tell if the train in the station is about to pull away by the descending, two-note warning sounded as the doors begin to close. It was created as a tactical sound, a warning to keep people from being pinched between the doors. Conductors have transformed that sound into a strategy. They often fake the door-close, playing the sound a few times before actually clicking the doors all the way shut and signaling to the driver to head on. They've turned that everyday tactical sound into a story that makes you feel something: anxiety, a rush, maybe even hope if you're midswipe at the turnstile and think you still have a chance of hopping onboard. If you're running late and already on a train that's been held at a station, that two-note chime can make you feel relief that you're finally getting under way. If the conductor fakes the door-close in that situation, the same sound can make you feel frustrated.

The way conductors have harnessed the power of a simple warning sound is nothing compared to the way sound is employed in transit systems in other countries. The Tokyo transit uses small jingles to signify certain stations, meaning you never have to use your eyes to know where you are. Moscow's metro uses a male voice for trains running clockwise around the ring line and a female voice for counterclockwise trains. In an instant, you know you're going the right way, and you feel reassured. These are what I refer to as functional sounds: sounds that give you very clear information that benefits your experience. In the process of soundscaping, these will rise to the top naturally and you'll start to actively listen for them, because they pack so much meaning into such a little space.

Here's another. Boot up your computer. If you're the owner of one of the more than two hundred million Macs sold, what's the first thing you hear when you've successfully started or restarted your machine? On a basic level, the sound you hear tells you you've held down the power button long enough to get things going. You might not consciously notice the sound each time you power up your Mac, but the fact that it's there means you don't even have to look at your computer to know it's working right. It's not just executional feed-back. What does that sound make you feel? Refreshed? At ease? Comforted, even? On your way to productivity? This is what I refer to as a brand-navigation sound. It's a branded, ownable sound — it could only be a Mac — that is both functional *and* emotional and gives you a lot of valuable information in just a few seconds.

Apple's start-up sound is a fun example, because it wasn't always so Zen. For a while, the sound of Apple was the sound of something going horribly wrong. Starting up a Macintosh began with a com-bination of notes that early-eighteenth-century music theorists and composers called the devil's interval — a tritone. It's any two tones that are three whole steps apart and played at the same time, like middle C plus the F# above it. It's disconcerting, provoking a feeling of agitation and anxiety. So irritating was this combination of notes that tritones were thought in Gregorian times to invoke evil incar-nate. Tritones were all but banned in early religious music. And yet, there was a loud tritone, kicking off your experience with an early Macintosh. It wasn't the experience a customer wanted. It wasn't what Apple wanted to give them either.

Jim Reekes is the guy who spotted this problem. He's a big rea-son why you love your Mac, even after it crashes, and the first Macs crashed a lot. He's the son of an early Apple employee and an infor-mal student of all sorts of uses of sound who as a young man found himself haphazardly chasing parallel careers in sound and comput-ers. In the mid-1980s, when Reekes was in his twenties, he'd already built an Apple computer from spare parts that he got from his dad. He'd also built his own synthesizer.

In 1988, his tinkering with Apple machines landed him a job as an engineer with the company. Steve Jobs had been ousted from Apple three years earlier. The visionary cofounder would come back to res-cue his company in 1996 and return it to profitability by 1998, but Jim started his twelve-year career at Apple during a rudderless period that he calls a "turd sandwich."

Reekes describes sound back then as "yet another fucked-up

project at Apple." There were a whole host of problems with sound, not the least of which was how it . . . sounded. "One of the things I wanted to do was replace all of the old sounds," he says. He recorded his own car alarm for one proposed app that never got picked up. He recorded a coworker saying *quack* for the famous sound that made its way to early Macs. When he began rewriting the computer's sound manager, he winced every time he heard that start-up sound. It had been devised by a highly educated mathematician on staff. Reekes, by contrast, dropped out of college in his senior year and kicked and scratched his way into the upper echelons of the computer industry. Reekes became determined to take over Apple's sonic experience. "It's not just me that thinks it's bad. It's bad," he says of the sound he sought to supplant, the tritone. "It's been bad throughout history. It's literally the most dissonant sound you can make."

He admits that the sound had played so much that people started to forget how bothersome it was. But whether you realize it or not, your computer's start-up sound frames the experience you go on to have with it. It's a symbol of what's to come — Reekes calls it an ear-con (analogous to an icon). This tiny sound leads off all the connections that your computer enables — to other machines, to worlds of data and knowledge, to people. When a sound like this is played in the right situation at the right time, it's incredibly potent. Reekes set out to design a Mac sound that would spark a magical experience.

"I thought, I gotta have this meditative sound," Reekes says. "I used to joke about it being a palate cleanser for the ears." He had to design it to fit a lot of different machines (Apple was considering many versions back then) and all of the various configurations where the sound would play — tiny speakers on the cheapest line of Macs, the beefier sound coming out of the (then) new Quadra series, even professional speakers in actual music studios hooked up to Macs. He ended up with a big two-handed C-major chord. It's in stereo. It fades back and forth, left to right. There's a bit of reverb in it. It's played by a bunch of string sounds and even what Jim describes as a "chiffy" bamboo flute sound. "It's a calm sound. And I knew that people understood C major, even nonmusicians. And it'd still feel interesting to people who are in very good studios. I was trying to reach a very broad audience with the intent and type of emotion I was trying to evoke."

When Reekes put it on a few of the early prototype machines, his superiors balked. It's a common reaction from people who don't quite get the magnitude of the opportunity in sound — there are still a lot of innovative people who need convincing, even though they feel the impact of sound daily. "You don't know what you're asking for," Reekes's bosses told him. "No one would let me change it," he says. "So I had to sneak it in at night." He went into the office in the wee hours, changed the code, inserted his sound, and eventually enlisted the support of one of his superiors, who looked the other way when others protested about the sound change. In the end, the computers shipped with Reekes's sound. It was a coup at a company that's since become known for its iron grip on design.

The Macintosh Quadra 700 came out in 1991. The reviewer of the machine in the now defunct computer bible *Byte* magazine wrote: "I knew I was in for something great when I heard it turn on."

"I'm like, 'Exactly! Victory!'" Reekes says. "That's exactly what I was trying to do!"

Lots of people at Apple subsequently tried to change Jim's start-up sound, he says, and he always argued against it. "It's like a logo, you don't keep changing it! Change isn't bad, it's just that it needs to be better." Although no one has definite proof of this, Jim believes Steve Jobs himself finally fended off any alterations to it when he came back to Apple in 1996. Jim innately understood that this particular sound was a *strategic* imperative — it wasn't just a tactical decision. It had to project the Apple brand personality, and because of consistent use over generations of the product, the sound is a lasting symbol of Apple's "think different" philosophy. It's synonymous with the entire product experience.

The earconic start-up sound proved the power of sound in a computing experience. That sound has remained essentially unchanged since then. Only minor tweaks have been made, despite numerous operating system and feature alterations, lots of new hardware, and tons of icon and font changes. No matter what Apple innovations come up, the start-up sound stays mostly the same and brings its customers the same satisfying *bong* when they first turn on their new Macs. After the inception of the start-up sound, engineers and designers started to pay more attention to how sound was used in Apple products. He's rarely credited by name, but Reekes's renovation of the sound manager in early Macs enabled all sorts of subsequent functional, delightful sounds. Think about the gratifying, refreshing *whoosh* you associate with sending an e-mail. Apple couldn't patent e-mail, but it owns the sound millions of users associate with a sent message and makes sure it shows up on every e-mail-capable Apple device. And now there are probably no hearing humans in this iPod and iPhone generation who haven't tapped the virtually limitless library of music in their pockets to consciously or unconsciously change their moods during a stressful day. Instantly accessible music is how we convert intolerable chores into enjoyable escapes — a trip to a packed grocery store is far less irritable with earbuds in. We use music to create a virtual space bubble on a crowded bus or subway or in a waiting room. Even if there is other music present, many of us prefer to replace it. These days, we're already taking steps to score the soundtracks of our lives.

Even without technology, you already know how to use simple sounds to tell stories, shape people's impressions, or just get what you want. What sounds do your shoes make, and how do you think about that when you go to an office? What do you do when you want to make a toast or signal your desire to say something important? How do you stop a child from touching a hot stove when he's out of your reach? And what do you do when you want to make all the sound go away? Shhh!

The growing need to control sound is the byproduct of a world constantly teetering on the edge of sensory overload. That's also why silence is, more than ever, one of the most essential aspects of a modern sonic strategy. When we create a desired experience, all we really have to work with is sound and silence — just as visual artists fundamentally have only positive and negative spaces for their designs.

Silence — or perceived silence — is better than sound with no emotion or purpose. The same autonomic system that keeps you from being distracted by your own breathing also allows you to ignore another set of sounds: sounds that don't ring true, benefit you, or help you understand how to get what you need in a particular moment. I call those sounds sonic trash. They're not just random noise. Sonic trash is the wrong sound, or the right sound telling the wrong story at the wrong time. The identifying characteristic of sonic trash is that it always amounts to a missed opportunity — to tell a story, provide meaning, or make someone feel something. Anyone who ever heard the ear-piercing crackle of the short-lived SunChips "green" bag probably remembers first and foremost the disturbing noise it made, not that it was remarkably biodegradable. Almost the only thing the media talked about was how loud it was, and when Frito-Lay saw a massive dip in sales, it pulled the bag from production. Too often sonic trash is ill-considered jingles, insistent music, beeps, or dings that get bolted onto brands, stories, or experiences. Lots of car brands do this (I'm looking at you, Nissan). Some blare at precisely the *wrong* moments, like candy wrappers at an opera. More often, sound is used to fill gaps when what's needed most is silence — wireless Bluetooth speakers that constantly beep as they connect and disconnect with your devices throughout the day; a dishwasher that plays a little jingle at 4:00 a.m. when it finishes the load you put in right before you went to bed. Sonic white spaces — effective gaps of silence — can prepare you for emotions just as powerfully as sound does. They can heighten the drama of a film or television score at the edge of an expected climax. Humans yearn for contrasts in their soundscape; they provide moments of interest and rest, tension and release.

Disney is a master at balancing sound and silence in its theme parks. There, complex arrays of speakers and finely tuned environmental sounds kick off subconscious stories that start the moment you open your car door in the parking lot. "Sound is a place-setter. It's the mood-setter," says Joe Herrington, principal media designer for the Disney parks at Walt Disney Imagineering; he's been there thirty-three years. "It puts you in the right mindset right off the bat. That of course is working in conjunction with the colors that you see, the costumes that you see. Immediately all of those things are working together."

When all of this comes together just right, an amazing thing happens to the sound itself: It disappears. You don't notice it as a separate component — you notice its impact only as part of the larger experience.

Eight days after hearing the world anew — that thunderous door slam, the bone-rattling crunch of her own chewing, and, for the first time, her daughter's voice on the phone saying "I love you" — Sarah Churman went on Ellen DeGeneres's show to talk about the experience. She described "the rain, thunder, birds, things that normal people wouldn't think about hearing, my husband snoring, myself laughing."

At the end of the segment, Ellen sprang a surprise on Sarah and her family: Envoy Medical Corporation, makers of her hearing device, would pay for the procedure and equipment to "turn on" her second ear. Then Ellen presented Sarah with an oversize check from Envoy made out to Sarah for thirty thousand dollars so she could reimburse her mother-in-law, Lari, for the retirement savings she'd spent on Sarah's first aural implant.

On March 9, 2012, a little more than five months after Sarah began to really hear for the first time, she and Sloan made a second video and posted it on YouTube. In it, the technician asks about life with a single implant and whether it seemed odd not being able to hear "in surround sound." Everything about hearing had seemed odd to Sarah at first. She didn't have much of a concept of sound localization, and she says she still asks people where, for example, a siren is coming from when she hears it. Hearing through one ear is a lot like seeing through one eye — it makes it hard to pinpoint relative location. As Sarah's second implant comes online, her face lights up with wonder again. "Oh that's weird," she says. "I don't know how to explain it. . . . It's loud all over again. I didn't expect that. I can't wipe the grin off my face."

The technician asks Sarah's husband, Sloan, to speak so she can fine-tune the volume. Sloan makes a few sounds. The technician makes some adjustments. And when it's finally all about right, the tech asks Sloan to test it one more time. Sloan says he probably picked up the little joke he made "subliminally" from the bespectacled pitchman in the ubiquitous Verizon wireless commercials aired heavily in the early 2000s:

"Can you hear me now?" he asks.

Sarah could hear more than a whole new world of sound; for the first time, she could tell where those sounds were coming from and get a sense of how far away they were. Can you tell how fast someone is coming up behind you or how far away an approaching siren or semitrailer is? How? Sound. At twenty-nine, Sarah was finally experiencing a version of the natural sense that's been a big factor in the survival of our species. Sound localization is critical to help us determine friend or foe and make decisions about fight or flight.

In the months after she gained hearing in both ears, Sarah found herself in a unique position to discover how sound transforms all kinds of otherwise mundane happenings into emotional experiences. She realized how often sound came before sight and how much of a story it could tell. Sound could even draw her attention to smells, tastes, or touches she hadn't noticed were there. "It's just made me grateful — for everything in life, grateful to be able to experience things even at the age of twenty-nine like a child," Sarah said of her new sense of hearing. She still had a lot to look forward to. So do you.

2

The Boom Moment

REMEMBER THE RESTAURANT from the introduction that serves the sizzling fajitas? Did you guess that we were at Chili's? As of 2013, at fifteen hundred locations across every state, thirty-three countries, and two territories worldwide, Chili's sells 60.4 million pounds of fajita meat per year, "four times the weight of an average U.S. military submarine," according to Brinker International, Chili's parent company. In this chapter, I'll make the case that no ingredient was more important than sound in making Chili's synonymous with sizzling fajitas.

First off, Chili's didn't invent fajitas. Far from it. That credit would more likely go to Juan Antonio "Sonny" Falcon, who named the belt of meat that surrounds the midsection of a cow. He called it the *faja*, Spanish for "belt" or "sash." It was mostly considered trimmings. But Sonny, whose family owned a meat market, came up with a way to season and grill those tough cuts so they'd be tender and tasty. In 1969, at the Dieciséis de Septiembre celebration in Kyle, Texas, in the Rio Grande Valley, he opened his first booth to the public. At a time when most catered events served food at a counter and put the cooks in back, mostly out of sight, "I made it a point to set my grill right up front," Sonny says. "And they could see exactly what I was doing." He served his meat on flour tortillas — no accouterments, hot sauce optional. He called his simple dish *fajitas*. They were a huge hit wherever he went. A local paper dubbed him the Fajita King. But soon restaurants caught on to his sizzling-steak concept and brought it indoors.

Chili's wasn't the first restaurant to adapt Falcon's dish either. In the 1980s, a few restaurants in Texas were gaining a reputation for fajitas served on a sizzling platter. The Round Up in Pharr, Texas, comes up in fajita lore. So does On the Border in Dallas. The restaurant at the Hyatt Regency hotel in Austin, Texas, is most often mentioned as a pioneer in sizzling-fajita history. George Weidmann, the hotel's German-born executive chef, debuted a wildly popular gourmet version of sizzling fajitas in 1982 at the hotel's restaurant, La Vista. The spot overlooked Town Lake and had a flowing creek, called Branchwater, running through the middle of its bar, where patrons sampled experimental Southwest-themed spicy martinis and Mexican margaritas. At the restaurant's peak, La Vista's 182-seat dining room served a thousand to twelve hundred customers a night. By 1984, Weidmann and La Vista produced more than thirteen thousand orders of fajitas a month — they made up 80 percent of all dishes served at the restaurant. Weidmann did a fancier version of the fajitas you probably know today. It featured a tender, prime cut of beef, seasoned with his own blend of herbs and spices, and it was served with flour tortillas plus peppers, onions, guacamole, and the house salsa, all neatly arranged on a white-hot platter that sizzled and smoked as it passed by guests. "You could hear the sizzle, and the smell permeated the atrium dining room," says Lance Stumpf, who succeeded Weidmann (who died in 2001) as La Vista's executive chef and is now general manager of the Austin Hyatt. "People recognized it as 'Ooh! Something hot is coming up behind me!'"

La Vista was one of the first places to stumble upon sound in the fajita recipe. Weidmann, however, was convinced that it was his unique choice of spices that made the dish a success. That became a roadblock when Hyatt tried to replicate the dish at other restaurants in the hotel chain — they couldn't do Weidmann's spice blend in bulk. That, Stumpf says, is why sizzling fajitas never became Hyatt's signature dish. But that's the chef in him talking. What makes more sense is that Weidmann put the emphasis on the wrong ingredient.

You don't remember Chili's fajitas because of their spice blend or cut of beef — Chili's restaurants sell plenty of chicken and vegetarian fajitas too. In 1984, as the Hyatt struggled to modify its fajitas to fit in other hotel restaurants, Chili's rolled out its version in twenty-three of its locations, with great success. People lined up around the block waiting to get a table. Long before the term became popular, Chili's sizzling fajitas went *viral*. Cooks called it the "fajita effect": When the first order of the night came into the kitchen, the cooks fired up several skillets and started prepping ingredients for the bunch of orders that always came soon after. The boom moment of the first sizzle of the night always kicked off a multisensory chain reaction that made the whole dining room want the dish. In the wake of the massive success of its sizzling fajitas debut, Chili's printed up T-shirts with the message *I survived the Summer of Fajita Madness!*

The restaurant chain didn't do fajitas first or best, but it did them loudest. It even put the sizzling sound in its first TV ad. While other restaurants tried to make fajitas that tasted better or looked fancier than the competition's, Chili's kept the recipe simple and relied primarily on the sizzle to not only turn heads but also trigger a barrage of senses: you hear them, then you notice the smoke and smell the aroma. Add up all of that, and you don't just taste or see a neat-looking dish with Chili's sizzling fajitas. You *experience* it.

No other sensory input is more efficient than sound in helping craft these kinds of experiences. The right sound at the right time has the power to tell a rich story. Without your even realizing it, sound triggers memories and emotions. It makes you *feel* something instantly. When that happens, the results are bigger than sales numbers or effective marketing.

That's when you get a boom moment. Boom moments happen when a sound triggers this kind of multisensory experience — a complex mix of memories and expectations wrapped in feelings that aren't immediately explained by the sound itself. If I played you the sound of sizzling fajitas, then told you it wasn't, in fact, meat hitting a white-hot skillet but rather a person burning his or her hand on a hot stove, or water from a firefighter's hose landing on the flaming roof of a home, you would react with a completely different set of feelings. The sizzle alone isn't what's so distinct; it's the power of that sound to surprise and delight your ear in an unexpected setting, then usher you through the rest of the sensory experience that naturally follows. This chapter will show you how Chili's and others put the powerful emotional impact of sound to work to create experiences you remember. It's about discovering which sounds at which instants make for boom moments. You'll start to see how to spot them when companies or individuals pull them off effectively, and you'll begin to understand how to create them for yourself or your company. You'll also see how cultural milestones can become boom moments, points at which something embeds itself in our collective consciousness, often with a common set of feelings or experiences that help us form bonds with the experience and with one another.

To understand the mechanics of boom moments like the one Chili's discovered, you need to understand how deeply you're wired for sound. Your brain is constantly listening, whether or not you are conscious of it. It craves patterns, and then it craves exceptions to those patterns. Sound is incredibly effective for feeding these brain cravings. The experience of sizzling fajitas starts with the sound of liquid spilling on a white-hot surface (that's precisely what happens). It's a sound that pops in the sonic environment of a casual-dining restaurant. This is the same kind of sonic *pop* that seized the attention of early humans in their travels through their environment. They tensed up, breathed quicker, and got a jolt of adrenaline when they heard a rustling in the bushes that signaled either an impending attack or an opportunity to attack their next meal — the caveman equivalent of sizzling fajitas.

The experience often begins with a startle, which is a distinct set of automatic, neurological reactions to stimuli. Only sound, touch, or a loss of balance can cause a real full-body startle (think about what happens if you hear someone shout "Boo!," if someone sneaks up and touches you on the back, or if you slip on an unexpectedly wet floor). You can't become immune to a true startle, because it's a reflex. You might have a quick adverse reaction to a gory crime-scene photo or even want to hightail it out of there if you see a snake slither up next to you, but you can desensitize yourself to that stuff (otherwise, homicide detectives and zookeepers would be nervous wrecks). You can't be truly startled by things you see. It's not the sudden *appearance* of Jason Voorhees's hockey mask in *Friday the 13th* or Freddy Krueger's charred face in *A Nightmare on Elm Street* that makes you jump, it's the staccato strokes on strings or the ghoul's yell or the victim's scream. Don't believe it? Watch one of those movies — or, better still, the black-and-white shower scene in the original *Psycho* — with the sound down. In an interview for this book, film composer Hans Zimmer joked that when people get scared at horror movies and cover their eyes, they're doing the wrong thing: "They should be covering their ears!"

In the event of a startle, within ten milliseconds — thirty times faster than the blink of an eye — a five-synapse circuit causes you to jump, sharply shrug your shoulders, dip your head, and turn toward the source of the sound; this is known as the orienting response. Your heart rate and blood pressure spike too. In less than fifty milliseconds — still six times faster than the blink of an eye — you've identified the sound and where it's coming from. In the actual time it takes for you to blink, sonic input gets directed through your auditory cortex to other parts of your brain that control memories and emotions — the hippocampus and amygdala, to name a couple.

That's when a sound can become more than a sound.

Petr Janata, a neuroscientist and professor at the University of California, Davis, in the psychology department and the Center for Mind and Brain, has observed fMRI scans where the playing of music, even in the first second, corresponds with blood flow to parts of the brain that control premotor functions. Depending on what that sound turns out to be (an explosion, say, or the blast of the first few notes of a familiar tune), it can also correspond with blood flow to sections of the brain associated with cognition or sense of self. "The content of the memories or even visual areas presumably engage with reliving those mental images," Janata says.

Some sounds heard at times of particular relevance to the listener can activate sections of the brain responsible for deep, emotional, even visual stories. While no one's completely mapped out the full neurologic constellation behind most sound in the wild, and establishing causality is a trickier business than reading fMRI scans, it's clear that sound is a really efficient kind of input. The most efficient, in fact. Smell is a powerful trigger of memories and emotions and can even make you pay attention to other sensory input, but the brain's response to smell is much slower than its response to sound. And smell is much less developed than hearing in most people, and relatively unreliable. You can't use smell alone to triangulate a source at any significant distance. Vision activates large portions of your brain, and your eyes can make you pay attention to an object's associated smells or flavors — as anyone who's ever pulled up behind a garbage truck can attest — but only if you're already turned toward it. Plus, you can perceive only about twenty-five visual events per second. You can perceive two hundred auditory events in the same amount of time. This all helps explain why you react first to *hearing* sizzling fajitas. *Then* you see the steam or smell the frying onions and start to crave them.

But, neuroscience can't explain how Chili's came to own sizzling fajitas. People bring a bunch of environmental baggage into that scenario — expectations about the atmosphere, the menu, and the whole experience they associate with a Chili's. There's the physical, visual environment surrounding the fajitas presentation, details that might seem meaningless until they're organized by a boom moment. Then they become an inextricable part of memory rooted in and easily recalled by sound.

To further explain what's happening beyond the sizzling skillet, we

One-Second Brain Science

We react faster to sound than to all other stimuli.
→

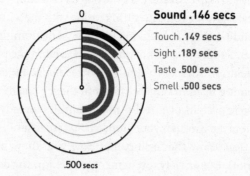

Sound .146 secs

Touch .149 secs
Sight .189 secs
Taste .500 secs
Smell .500 secs

Complex perceptions and reactions to specific sounds are often faster than reactions to other sensory input.

↓

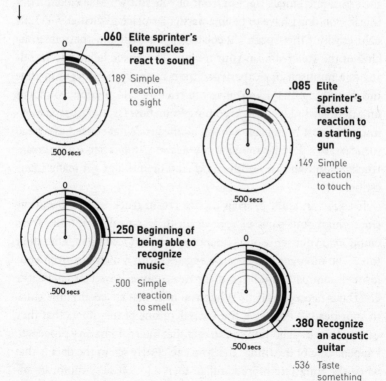

.060 Elite sprinter's leg muscles react to sound

.189 Simple reaction to sight

.085 Elite sprinter's fastest reaction to a starting gun

.149 Simple reaction to touch

.250 Beginning of being able to recognize music

.500 Simple reaction to smell

.380 Recognize an acoustic guitar

.536 Taste something sour

have to skip to dessert. It's the dog days of summer. You're beating the heat however you can: sitting inside with the windows open or the air conditioning on; lounging in the pool or running through the sprinkler. Suddenly, a tinkly tune bends around the block and grabs your attention. Almost instantly, you recognize it and can tell where it's coming from. It's the music from the ice cream truck. And whether you're a child or a full-grown adult, it's already sending butterflies fluttering through your stomach. You're in a full-on boom moment.

You know the feeling, and now you know how sound can rally emotion — anxiety, excitement, hope, joy, thrills — in your brain. It organizes the other senses to support this experience. The sound of sizzling fajitas in a restaurant triggers you to assemble a story from the sights and smells right in front of you. But with ice cream truck music, you don't have to be in a particular place to virtually feel the cold crack of the frozen chocolate-dipped shell and the refreshing chill of the ice cream on your tongue. The music helps you reconstruct a multisensory experience from memory. You begin to tune out everything that doesn't give you relevant information; you focus on where the tinkly music is coming from, how fast the truck is moving, and, most important, what it all means. What started as a head turn toward a familiar sound is now a *limited-time offer.* An ice cream truck becomes *an ice cream truck!* And then things get really interesting.

If you're an adult hearing the ice cream truck for the first time since childhood, parts of your brain that control everything from emotion to your sense of self to movement start feeding into the story too. "The moment you become engaged with a memory from your more distant past, you bring these areas online," says Petr Janata, the UC Davis neuroscientist. "When you have that ice cream tune stuck in your head — 'Pop Goes the Weasel' or one of the others that they use — just the suggestion can trigger that internal imagery process to happen. One of the things that you can clearly see in the data is that when the songs are more familiar, then a lot of these motor or so-called premotor brain areas became engaged." Within milliseconds, you're not only recognizing the tune and feeling emotional about it but also rehearsing what action to take to respond to it. A child hearing ice cream truck music might start mentally practicing an urgent request for money from mom or dad or loading up a catalog of promises to eat his vegetables at dinner and brush his teeth later and maybe even clean his room all weekend — *whatever it takes.*

There's a story in my family, told much too often, about the times the ice cream truck came jingling down my block in New Jersey. As a child, I'd get so excited that no amount of premotor rehearsal helped me get the words out. I'd sprint to my parents and be breathing so hard and trying so desperately to organize the whole operation that I couldn't make a sound.

"Ice . . . money [gasp, gasp] . . . please . . . *anything!*" I'd blurt, before bursting into tears of panic at the fear of missing out on a frozen treat.

In any case, it's the tune that transforms this thing you didn't even know you wanted into an immediate must-have. You don't need to see the truck or the images of ice cream splashed across its sides to know you want it. The feelings seem to come out of thin air. Ice cream truck vendors stumbled upon that power to conjure a powerful story with sound almost a century ago, during the Great Depression.

In the early 1920s, Henry Burt of Youngstown, Ohio, began delivering Good Humor bars — ice cream on a stick — from his trucks. They were outfitted with bells taken from a child's snow sled. "Burt's choice of small bobsled bells for his ice cream trucks was significant because it both recalled the familiar timbre of soda fountain automata and signed a *wintry* sound that conveyed ice cream's relieving coldness in hot weather," writes ethnomusicologist Daniel Neely, a widely recognized expert on ice cream truck music. During the Great Depression, Good Humor sold franchises for a down payment of one hundred dollars. Paul Hawkins opened Good Humor of California in 1929 and shortly thereafter replaced the bells with music — a Polish folk song called "Stodola Pumpa." "The choice of a single, simple melody was effective because it anchored an easily remembered sonic brand in the ears of consumers," Neely writes. Unless you happened to be working at a toy store or, say, Santa's workshop, a tinkly tune stood out in just about any sonic landscape. It gave the brain the kind of input it could latch onto — your mind always looks to complete the musical pattern, over and over. "It's very easily reinforced and rehearsed and pleasingly so in most cases," says Janata. "You can't do the same thing with a visual image. You could watch the same movie over and over again, in which case you'd have a very strong representation of that. But music's just much easier in that way."

Think about the last time you got a song stuck in your head on endless repeat. By the time you catch yourself humming it or replaying it in your brain for the hundredth time in a day, you may have started to hate it. It's a concept known as an earworm. It might be a song you wouldn't own up to loving in mixed company. It might linger in your mind a bit longer than you'd like after you realize it's there. But for the most part, earworms are your brain's way of putting an enjoyable score behind mundane daily tasks.

Many prominent cognitive psychologists have studied the phenomenon, including C. Phillip Beaman, Tim I. Williams, Andrea Halpern, and J. C. Bartlett. They've all found that, contrary to popular belief, most people reported liking the music that was stuck in their heads. "The idea that these songs are invariably obnoxious songs or painful jingles is a myth," writes Ira Hyman, another cognitive psychologist who has studied the earworm phenomenon as well as music and memory. Based on my experience with many hundreds of clients in my composing career and sonic business, and as a sound-obsessed adult, I'd take that a step further: You're happy to hear a song play over and over again — in your head or otherwise — as long as it connects you to a pleasurable experience, helping you get what you want, understand a story, or providing a welcome distraction. When you're trying to figure out how far away the truck is or how many seconds you have to beg money out of your parents, you can't get enough of the ice cream truck song. When, as an adult, the ice cream truck parks outside of your apartment while you're trying to nap, it's annoying as hell. An April 2007 survey by the Council on the Environment of New York City and the eTownPanel of Baruch College found neighbors' activities, car stereos, and police, fire, or ambulance sirens to be the top three sources of noise complaints. Ice cream trucks ranked thirteenth out of the twenty-four most bother-some sounds in the city (more annoying than the noise of bars and nightclubs, buses and subways, and even birds and insects, which were also on the list) — a uniquely New York testament to the power of sound.

By 1949, all sorts of ice cream trucks played all sorts of tunes, in-cluding "Strawberry Blonde," "Little Brown Jug," and "Sidewalks of New York." Today, they play everything from "Turkey in the Straw" to "Pop Goes the Weasel" to "Ice Cream," by Andre Nickatina.

But one business, Mister Softee, founded in 1956 in Philadelphia by brothers James and William Conway, played only one unique song after the brothers relocated to Runnemede, New Jersey, in 1958. The Conways commissioned the song from Les Waas, a songwriter at a Philadelphia ad agency. It's called, appropriately enough, "Mister Softee," and although you never hear the lyrics, the song actually has them:

> *Here comes Mister Softee*
> *The soft ice cream man.*
> *The creamiest, dreamiest soft ice cream,*
> *You get from Mister Softee.*
> *For a refreshing delight supreme*
> *Look for Mister Softee.*
> *My milkshakes and my sundaes*
> *And my cones are such a treat.*
> *Listen for my store on wheels*
> *Ding-a-ling down the street.*
> *The creamiest dreamiest soft ice cream,*
> *You get from Mister Softee.*

The plan was to incorporate the traditional sound of bells in the song and broadcast it as radio ads. But the Conways liked the theme so much they decided to play it from all the Mister Softee trucks. After all, those were its primary points of contact with customers. In doing so, they recognized key components of an effective boom moment: the unique song was used strategically and consistently, and it told a true story at *just the right time.*

But it's also childlike. It sounds like a jack-in-the-box or a music box even though it's not "Pop Goes the Weasel." Ice cream music expert Daniel Neely writes: "On the road, vendors hope their music's tune and timbre will link perception to memory and lead to nostalgia — for childhood, for sweetness as for the Main Street of American imaginary." The song, essentially, means *safety.* In the case of the ice cream truck, a tinkly, nursery-rhyme sound is what makes it okay to literally take candy from a stranger. As cognitive psychologist Ira Hyman says, "You're buying a lot more than just ice cream. You're buying your childhood."

The use of a custom song was the crowning achievement in the Conways' approach. It transformed Mister Softee in a way that no bell or borrowed tune ever could. Since Mister Softee owned its song, it could never be confused with someone else's brand or message. It couldn't own ice cream, but it could own a piece of what it meant to enjoy it — from the way the experience reverberated in your brain and the way it made you feel to the way it became embedded in your memory and instantly came flooding back, whether that happened the following summer or decades later. "It was so new and so very different for its time," says Jim Conway, the vice president of the Mister Softee ice cream truck company he inherited from his father and uncle. "It really made the Mister Softee truck stand out from its competition." The family company now oversees three hundred franchisees running seven hundred trucks in eighteen states. It's a private company and doesn't disclose its earnings, but Conway says the company does "several million" in annual sales.

"Today's best-known ice cream truck tune is probably Mister

Softee's," Neely says. For a larger part of the country than any other single modern ice cream truck operator, Mister Softee literally owns ice cream music. And if you own the sound, you can own the experience.

Whether it's coming from a Mister Softee truck or another ice cream truck, the music makes you forget about everything besides what the ice cream man is offering. Think about those times you heard the jingle from your living room or garage or backyard. Were you mere steps away from a freezer stocked full of ice cream, higher-quality stuff, butterfat-laden pints, quarts, or half gallons of Häagen-Dazs or Ben and Jerry's? Did the ice cream music ever make you want to mosey over and dish yourself up a bowl from your own stock? Or did it make you want to hightail it out the door in pursuit of, say, Mister Softee?

"You go for the truck," says Jim Conway. "Going to the fridge isn't an experience."

Ice cream purveyors aren't the only ones to capitalize on the influential power of sound. Even a small piece of music or a sound played at the right time can trigger a memory or an action. When you see the Twentieth Century Fox logo and hear the drumroll play before the feature presentation in a movie theater, you know that the main attraction is starting. You might notice the lights dim further. And you know it's time to wrap up any conversation. It might even make you think of the first film you saw on the big screen. You might notice the smell of popcorn that's been lingering since you sat down. And you might decide it's okay to go ahead and crack open the Junior Mints that are sitting in your lap.

In movies themselves, repeated sonic patterns can appear in everything from incidental music to full-fledged themes or songs or even iconic sound effects. If you hear them later, even decades later, they transport you back to the time and place where you first saw the film and call to mind all of the senses and emotions you experienced then. Think about the distinct laser or TIE-fighter or light-saber sounds in *Star Wars*. At the first mechanical gasp of breath, you know

Darth Vader's about to show up, even before you see his shiny black helmet. And you *feel* his evil — a strategically devised low rumble. (Conversely, in *Return of the Jedi*, Darth Vader's high-pitched wheeze when his helmet comes off as he rediscovers his humanity instantly makes you feel sympathy for his sinister character.)

In 2007, Ford executives found themselves faced with the challenge of trying to replicate this kind of cinematic, emotional feeling — with a car.

It was going on seven years since they'd put out a special edition of the Mustang. These small-batch models surround the whole brand with a halo. They remind people how they felt years ago when a neighbor or a cool kid in town or maybe they themselves got a Mustang and instantly turned heads (and scored dates). Every few years, Mustang cues up that nostalgia and maintains the brand's place in the market as the maker of the quintessential American rebel-mobile. And in 2008 (as in 2001), they did it by creating the perfect sound. But perhaps not the way you'd think. "There's an expectation from enthusiasts that we're going to give them some kind of specialty car," says Mustang noise and vibration engineer Shawn Carney, who's owned seven or eight (he's lost count) of them. "Those are the cars that people tend to salivate over." And heading into 2008, Ford and Mustang lovers were particularly ravenous.

Plenty of people on Internet message boards speculated that Ford would release a new version of the Boss or Mach One, high-performance editions from the 1960s, but Shawn knew better. Inside the confines of its R&D facility, Ford was quietly building another Bullitt Mustang, a fortieth-anniversary version of Steve McQueen's highland-green muscle car, which he famously drove in one of Hollywood's most iconic chase scenes in the 1968 movie *Bullitt*.

In a private session, Ford managers met to take the prototype for a spin. It was a new car, but it sported many of the minimalist visual signatures of the roughed-up 1968 McQueen machine — no chrome trim, no pony on the grille. But Shawn saw a big problem. No one focused on the sound, and it was a disaster in the making. The distinct sound doesn't come from the engine, exactly. It's in the exhaust "note"—the gurgle heard in the moments before Detective Frank Bullitt mashed the gas and sped after two hit men in a 440 Magnum V-8-powered Dodge Charger through the rolling hills of downtown San Francisco. "This isn't something we're going to be proud of," Shawn said after he heard the car. "This isn't something that's going to be thought of as paying tribute to *Bullitt*." The stakes were very high. The reputation of the brand was on the line.

"We had to come up with our own interpretation of what this Bullitt [car] should sound like," Shawn says. But first, he had to convince Ford's designers that, as great as the car looked, it wasn't complete without the sound, and it was not going to meet anyone's expectations. If the car didn't have the right sound to initiate the whole multisensory experience, all of that attention to detail would fall apart—it wouldn't *feel* right. Ford was in danger of blowing a critical opportunity to win over the next generation of brand ambassadors who'd make web clips of their Bullitt Mustangs growling and peeling out. (They're all over YouTube. Google 'em.) "It's so critical to what makes a Bullitt a Bullitt," Shawn says. They had to act very quickly.

So Shawn pulled aside Paul Randle, Mustang's chief engineer, and told him what he had in mind for the car's exhaust. "We had to show what the opportunities were," Shawn says. He took Paul down to meet a friend who owned a factory 2001 Bullitt, the first edition of the specialty car. Shawn had his buddy fire it up so Randle would hear the unique sound of the pipes and intake that he already had experienced in his years with Ford. He wanted his boss to feel the Bullitt's boom moment for himself. "Right away, he's like, 'Oh, we're doing it,'" Shawn says. "As soon as he heard it, he bought into it." That's usually all it takes. Make people feel the difference, and they're instantly hooked. Randle made sure Shawn and his team had the budget to get to work right away. And that's where things got really tricky.

In 2001, when Ford created the first new Bullitt Mustang, the team members (Shawn wasn't working on Mustang yet) did their best to match the sound of the real car to the sound of the one in the movie, and they did a fine job. But in 2008, the Mustang's body and power train had been completely redesigned. Shawn had to start from scratch. And that's when he realized he could never perfectly match the sound of this new car with the one in the film. "There's no factory Mustang that's ever sounded like that thing," Shawn says. He was chasing a Hollywood ghost.

It was impossible to replicate the sound of the car in the movie from a technical standpoint. First, when you think about the way the Mustang sounded in the *Bullitt* chase scene, you're not thinking about the sound of an actual car. It's a mix of that tricked-out movie car's sound and racecar sounds dubbed in by Foley artists. The best Ford engineers could do was find a comparable pitch and frequency and modulation to make the new car stand out from any other Mustang on the market, just like the film version of the Mustang did. "What we're trying to do is match someone's perception of what a Bullitt should sound like, knowing that we can't match exactly what's in the movie," Shawn says. Second, it was impossible to make a street-legal Mustang as loud as the movie version, he says. It'd violate noise ordinances as soon as it rolled off the assembly line. "When it comes to the Bullitt, the movie piece is inspiration. But then we have to deal with the physical realities of knowing what sounds were used in the movie that aren't possible to put on an actual car."

The solution involved an arduous and mind-blowingly geeky process that started with a computer sound simulation — a set of headphones with a paddle shifter, a steering wheel, and a gas pedal hooked up to a screen — that let designers hear what a car sounded like on a test drive in a virtual setting before they actually started welding. In the simulation, they could pull out the sound exhaust and hear just the intake, or vice versa. They could mix it with environmental sounds. Then back in the real world, engineers got to cutting and installing lengths of exhaust pipe, toying with air pressure and how it was distributed to the left and right exhaust pipes. The metal beneath a Mustang's chassis became Shawn Carney's orchestra.

The 2008 Mustang Bullitt was released to great fanfare. About sixty-four hundred of them were made, and all of them sold, even with a price tag of $31,075. A group of Bullitt owners from the International Mustang Bullitt Owners Club agreed that the sound of the car was key in making it a success — and in making a car that reminded them of the movie. Club member Greg Autry first bought a 2001 Bullitt. When Ford invited him to drive an early model of the '08 edition, he fell in love. He later sold his '01 and bought a used 2008, number 3,383. "When I bought it, it had after-market mufflers on it. So I actually had to go and find the original mufflers to put back on it," said Autry, fifty-two. "Because I like the sound so much of the original the way it is."

Another club member, Paul Rocha, forty-six, who owns two 2001s and a 2008, talked about buying his first Bullitt in 2001; designers labored over the exhaust note on that car too. "I'd been trying to work a deal on a Bullitt at a couple of different dealerships, and both times the car was in the showroom." Then he found a small-town dealer that had one out on the lot, "and they gave me the keys to it. And up until then I was playing hardball with the dealership, but as soon as this small-town dealership gave me the keys, and I sat in the car, and I started it up, immediately I thought, 'I gotta have this car.' It was just the sound, then everything about it. As soon as I started it up and I heard the car, I knew I had to have it."

As they discussed the Bullitt — both the movie car and the reissues — Rocha, Autry, and other Mustang aficionados were quick to describe the color, year, and model Mustangs they first owned and, in one person's case, how "I wore out the springs in the back seat." Part of the job of the Bullitt — and any special-edition Mustang — is to trigger memories of those earlier cars. Asked whether any new Bullitt sounded exactly like the movie car, Rocha said, "I don't think you're ever going to get anything modern day to sound like an old-school V-8, but this definitely does remind me of it."

Most agree that Shawn and Ford got the sound right and that it successfully called to mind the chase scene in that 1968 movie. "I think of the original movie as a Rolling Stones song," Shawn says. "You're never going to invent another Rolling Stones song, but some of the covers are pretty sweet."

The Bullitt uses sound to trigger a series of reactions — first you hear the Bullitt's distinctive exhaust note, then you start to notice everything that's different about the whole car. You might notice the paint job and think, *Is that car* green? You might notice the subtle lack of details and question for a second what model (or even make) you're looking at — there's no pony logo on the grille, no chrome trim; visual white space sets the car apart that way too. If you got inside for a ride, you'd immediately feel a different kind of rumble. "One of the comments that you get is that you can actually feel the sound. People say, 'I feel it, like, pounding in my chest, like you feel at a rock concert,'" Shawn says. By design, it rumbles at a lower register than a typical Mustang. In fact, there's a fine line Shawn and his team walked between making the car sound muscular and making it so loud it was annoying to drive. They call it "candy." "If you get too much of it, it gives you a tummy ache," Shawn says. You don't need to hear the intense rumble constantly to realize you're in a beast of a machine. It's like hearing the Mister Softee truck song over and over outside of your apartment when you've had enough cheap ice cream.

Here are the key parts of the boom moment Shawn helped Ford create with the 2008 Bullitt: First, he came up with a sound that would jump from the background to the foreground. It just reached out and grabbed your attention. If you care even a little bit about cars, you can't help but soundscape this unusual exhaust note to the foreground of sounds competing for your attention. *What's that?* you might think, and then you turn your head. You're intrigued, and then all of the other stimuli start to fall into place — the lack of Ford or Mustang logos (visual white space) deepen your curiosity about this car until a whole set of other cues start to help you piece it together: the thick gas cap, the green paint, the wheels. That throaty, gurgly sound plays a vital role in assembling all of that visual evidence, a distinct power associated with a boom moment. Then it helps you recall an iconic car in a classic movie. From there, it's a short leap to all of the ways that that film and its car made you feel. "I'm trying to elicit emotions," Shawn says, and since Ford couldn't build an exact replica of the 1968 Bullitt, sound was the most efficient and effective tool to re-create the feel of the movie car. "I'm trying to give people a visceral experience. You're making something new, and it's okay to be a little bit different from that historical reference that everyone has in their memory. Blend them together and you have an experience."

What do an ice cream truck and a muscle car have to do with Chili's sizzling fajitas? Everything. The savvy designers, engineers, and regular entrepreneurs behind these sounds all created textbook boom moments. Ice cream truck operators and Mustang engineers did it deliberately, and Chili's got lucky and discovered the secret sonic sauce.

They started with a head turn. Tinkly ice cream truck music cuts through the atmosphere on a hot summer day, even if you can't immediately tell where it's coming from — just like sizzling fajitas cutting through the chatter and overhead music in a crowded casual-dining restaurant. The bark of the pipes on the 2008 Bullitt Mustang cause the same kind of head turn — because the sound rips through traffic noise at a frequency that's bolder than the average Hyundai (or even the average Mustang).

Then all of this sets off a multisensory chain reaction, one of the most important parts of the boom moment. In the blink of an eye, these sounds pull in sight, smell, taste, and the way something feels to help you figure out what's going on and what you should do with this information. In the case of the ice cream truck, it comes from memory — the childlike tune calls to mind the bright pastel-colored frosty treats and the feeling of something cold hitting your tongue on a hot day. The Bullitt Mustang's exhaust note lets you make sense of things that are already right in front of your face, meant to help you remember how you felt when you first saw Steve McQueen ripping up the roads and flattening the hills of San Francisco in his '68 Mustang.

Similarly, the sound of sizzling makes sense of the smells in the air and sights in the décor of a Southwest-themed casual-dining restaurant. I bet the mere mention of the experience in this chapter made you virtually smell and taste fajitas again.

Chili's founder Larry Lavine didn't order up a bunch of fMRI images before launching his sizzling fajitas in 1984. But having opened his first Chili's on March 13, 1975, in Dallas, Texas, he'd learned a thing or two about what worked in a dining room. More important, he recognized a boom moment when he stumbled upon it. He'd eaten silent versions of fajitas for years at local restaurants but says he never thought much of the dish, because "they didn't put them on a sizzling platter. They had a tiger by the tail and they didn't know it." The first time he really *heard* fajitas was at a neighboring Dallas restaurant called On the Border (which Chili's parent company would later own). "They were the ones that changed it and put it on a sizzling platter," Lavine says, and he quickly realized how the sound transformed the experience. "That made the concept." In the great tradition of restaurants borrowing from one another, he decided to add a simple version of the dish to Chili's menu. No fancy spices. No high-grade cuts of meat. Just a lot of noise. Right away, he says, "we could see the difference the sizzle made."

The rest is history, the "summer of fajita madness."

"It's hard to come up with that kind of invention in restaurants," Lavine says — at sixty-eight, he's developing a barbecue-concept restaurant with a "crackling" fire pit in the middle, where the barbecue is cooked. "There are not many food items that you bring out that make sound, so it really stands out. . . . You have the smell and sight *and* sound. If it goes through the dining room, somebody's going to ask, 'What's that?'"

So why do you think of Chili's when you think of fajitas? To paraphrase the old marketing slogan, Chili's didn't sell the steak, it sold the sizzle.

3

Sonic Landscapes

Bᴀᴄᴋ ɪɴ 1999, when director Frank Darabont was shooting *The Green Mile,* his costume designer went to world-famous cobbler Pasquale Fabrizio, whose family has run Pasquale Shoe Repair in Los Angeles for more than fifty years, with a problem. The film is set in the 1930s in the death-row wing of a prison. Tom Hanks stars as a guard. And the actor was outfitted with a historically accurate pair of vintage work boots. Thing is, during a somber scene when Hanks was walking the corridor — the green mile itself — leading to the electric chair where he would carry out an execution, his boots were squeaking. It was distracting the director. So, according to Pasquale, the costume designer sent him the boots and asked him to take out the squeak. It's a common problem in old leather footwear, Pasquale says. There's a steel structure in the middle of each boot that gets rusty or bent or goes loose against the leather. That's where the sound comes from. He took it all apart, replaced the guts in the boots, and sent them back to the set.

A few months later, the same boots came back to Pasquale's shop. The director was doing reshoots. And he wanted the squeak back in.

"It was more dramatic to hear the squeak," Pasquale says. The default in a scene like this is to get rid of all of the environmental sounds to keep the focus on the story. While Darabont tells me he doesn't remember the specifics from the filming of *The Green Mile*, he says that taking out a squeak from a pair of boots in a quiet scene sounds like something he'd do. "Having a recurring sound like that can be wildly distracting and create a huge amount of work for the sound editors in post when the director says, 'Fellas, you have to lose all those damn squeaks!'" He doubts the part about returning the shoes to the cobbler to have the squeak put back in. "It seems to me I'd rather add something like that in post, because it's far more precise in its use that way," Darabont says. "At least that's how I'd handle it today." Darabont does remember sound playing a powerful role in the film in other ways, though. "I wanted a very distinctive and powerful 'electricity' sound for those moments in the film when the switch is thrown on the electric chair during an execution," he says. "I wanted it to sound very elemental and scary, as if a roaring, voracious beast had been unleashed. And in addition to the bass-heavy bottom layers of the sound, I also wanted a high-pitched element that I described to my sound designer as a 'screaming demon' in the electricity flow. If you listen to that sound on a great system, you'll hear all those layers brilliantly woven into it. And, yes, you'll hear the demon's shrill, high-pitched scream in the current."

Whatever the specifics, Darabont says he harnessed the power of sound to heighten the drama in the story. And you don't have to get anywhere near an electric chair to experience the power of small sounds every day, and not just from shoes or physical things, but from people. Who's the person in your office or class or life that has a funny sneeze? Maybe it's a frustrated, booming *achoo!* or a big, breathy setup followed by a cartoonishly apologetic *ah-chyeeew?* Each sneeze tells a story about its owner — she's bold or brash; he's docile or self-conscious. Think about someone you know with an infectious laugh. Now think of Seth Rogen's laugh. It helps define his characters, but if he was sitting behind you on a long cross-country flight, it might get on your nerves.

Sound guides us every moment of every day. In many interactions, we gather troves of sonic input about people or things long before we form opinions about appearances. The information we get from sounds, even the smallest ones, combines with emotions those sounds make us feel or past experiences they make us recall. Each of us is constantly calculating all of this at breakneck speed and using it to make choices about whom to talk to, listen to, empathize with, trust, ask for help, or even sit next to on a bus.

Even with all this automatic sound processing, you pay conscious attention to only a fraction of the sounds that are around you all of the time. After you read this, be prepared to notice a lot more of them. Your life might get a little louder at first. But becoming aware of more of these sounds is just the first step. You'll also learn to better curate the most useful sounds and discard the meaningless ones.

Start, like Frank Darabont and Pasquale Fabrizio did, by thinking about the sound under your feet. You take it for granted when it's aligned with what you expect — a solid thunk of concrete, a hollow bang of a wood floor, or the absence of sound on a plush carpet. In an age where people are more likely to be staring at screens while they walk rather than at the ground, the sound (and feeling, in the form of some of the same vibrations that make up sound) of every footstep helps them determine when to pay attention to where they're walking. If a certain vibration of the surface under your feet suddenly gives way to silence or a softer sound, well, you've probably just stepped in something. In New York City, for example, the sudden reverberating clang of metal beneath your shoes can make you quicken your step or hop a little — you'll quickly realize you've stepped on someone's metal basement hatch, and most New Yorkers have heard horror stories of these doors giving out and sending pedestrians plunging down a dark hole. Or late at night, the pace of footsteps behind you in a dark alley can lead you to worry that someone's about to rob or assault you. You speed up and listen to hear if the follower's footsteps do too, even if you don't dare look back.

In modern life, you get a whole range of important and trivial stories from functional sounds like these, to say nothing of the sounds you encounter during interactions with technology and machinery.

Your car reminds you to fasten your seat belt with a simple — and annoying — beeping sound. Those sounds result in a 3 to 4 percent increase in seat-belt usage, according to a 2009 National Highway Traffic Safety Administration study, and "systems with more aggressive reminder displays and more frequent repetition patterns were perceived to be the most effective, and sounds were perceived to be more effective than visual displays." Or think about a buzz in your pocket from your smartphone during a meeting telling you your wife has texted you. The determined click of high heels coming down your office hallway reassures you that you beat your boss to the office (but she might be heading your way to grill you). You don't have to seek out or even actively pay attention to these highly informative sounds to use them to your advantage.

"Iraq was the first time I noticed it," says U.S. Army lieutenant colonel Robert Bateman, a military historian who's taught at West Point Academy and did tours in Iraq and Afghanistan. During his first tour, he lived in downtown Baghdad. Sound would bounce off buildings and down alleys, but he gathered a tremendous wealth of information by listening carefully to the details of what he heard. He could tell who was fighting, who was winning, and when the conflict was about to end; he could pinpoint the source of a battle or explosion to within fifteen or twenty degrees. The sound of the gunfire provided information. "You can tell when a unit is running out of ammo, because their rate of fire slows," he says. The American weapons themselves had distinctive sounds; the M4, M16, A2, and M249 are 5.56 mm weapons, meaning they have relatively high-powered charges but small diameters. "They have a sharper sound," Bateman says. By contrast, the venerable AK-47, used by insurgents as well as the Iraq and Afghan militaries, is a 7.62 mm weapon. "It's bigger around, and a deeper or throatier round," he says.

In addition to the sound of the gunfire, the frequency and patterns of it told Bateman stories. If, for example, an American army unit got hit with an improvised explosive device, he'd hear the boom . . . then silence . . . and then maybe, after a while, a burst of gunfire, but not always. If an Iraqi special police battalion was hit, he says, he'd hear the explosion, then mostly silence, then some sustained fire. "But if an Iraqi army unit was hit — their discipline tended to be lower, their enthusiasm for gunfire tended to be higher — then they would do what we called the 'death blossom'" (a reference to the movie *The Last Starfighter*), Bateman says. "Every man would fire his entire magazine of ammunition randomly outward from his perimeter. You could tell what had just happened from kilometers away before any reports were sent."

Combat sound is the kind of thing you learn only as a soldier on the battlefield, he adds. It's a different experience from training exercises, when you're going up against fellow troops with similar types of guns and ammo and a similar set of tactics. Sound is not a part of any formal training he knows of, but it has been an essential part of battlefield warfare since the days of clubs and stones.

Military sonic intelligence isn't limited to the interpretation of external sounds. In 1944, the U.S. Army's 3132 Signal Service Company Special worked with engineers from Bell Labs to record infantry units at Fort Knox. Then they mixed those sounds together and broadcast them through huge speaker arrays on the battlefield to deceive the German army into believing that fighting was happening where it wasn't or that forces were as large as three hundred thousand troops when in fact they were much smaller. The sounds could be heard as far as fifteen miles away, and they were used during the invasion of Normandy and through the end of the war. The eleven-hundred-troop outfit that harnessed these powers of sound (plus several visual and theatrical forms of deception) was known as the Ghost Army.

You don't have to be staring down the barrel of a gun to appreciate the power of sound, though. It's the switch that flips on the human neurological autopilot system, whether it's a fight-or-flight situation or an everyday decision-making experience. Sound is the operator's manual of the subconscious, providing the cues that guide behavior in all sorts of situations.

Retailers and restaurants have started to examine how sound influences their business, paying attention to how it affects their customers as well as their employees. A Woodland Hills, California, company called Prescriptive Music curates custom playlists for a growing number of restaurants and chains, including the national group Umami Burger, whose spots you may have seen (or heard — the music they play rocks). Bill Chait, the force behind LA's Short Order, Playa, and other restaurants, believes music is a key early ingredient in the experience his establishments offer. "People consider the music a demonstration of whether this place is for them," he told the *Los Angeles Times* in a May 19, 2012, story. People associate the music in the restaurants they choose to visit with their own sense of identity, and the employees there do the same. Chefs have long harnessed the motivational power of sound in their kitchens. Culinary luminary Jesse Schenker of New York City hot spot Recette is famous for blasting Tool and Metallica while he cooks — it's practically an ingredient in his dishes. And servers placing orders know that they interrupt his headbanging sessions at their own peril. City Winery is another Manhattan spot where music and food mingle directly. In 2012, WXRT radio program director Norm Winer teamed up with City Winery's beverage director Rachel Driver Speckan for a successful event called Twenty Songs, Twenty Wines. They paired the drinks with the tunes and used the matchup to talk about inspiration. *Music taste* is rarely so literally interpreted. And celebrity chef and protopunk devotee Anthony Bourdain (whose hit TV show *No Reservations* my company and I were fortunate enough to score) stated in his book *Kitchen Confidential* that the playing of a single Billy Joel song in his kitchen was a firing offense.

Sound can, in fact, affect our perceptions of flavor. Unilever is the Anglo-Dutch multinational corporation that owns four hundred brands, including such beloved names as Dove, Lipton, PG Tips, and Ben and Jerry's. In 2012, it partnered with the University of Manchester in England to study the effect of noise on taste. The study found that people enjoyed their food more when they also enjoyed the background music playing. Loud white noise played at eighty decibels (about as loud as a blender or a washing machine), however, dulled the perception of flavor. When it's there, people perceive salt or sugar less intensely. What's even more interesting is that when that same background noise is going, people perceive more crunchiness. The study offers all kinds of insights about matching sounds and culinary objectives. Use white noise when you want to play up crunchiness, for example, and make sure you bring the beats and melodies if you want diners to appreciate subtle or complex flavors. Sound, in other words, is a vital yet often overlooked ingredient in meals.

There's a real-life situation where you've probably experienced this. How do you feel about airline food? It's pretty bland, right? While the low humidity in an airplane cabin and the cabin pressure affects the way you perceive taste, the Unilever study suggests that the lack of flavor in onboard meals can be partially blamed on the dull drone of the airplane engines. That type of noise makes you less sensitive to salt, sugar, and spices. But you do notice more crunchiness, according to the study, which would help explain why you're likely to pass on the in-flight Salisbury steak but ask for a second bag of peanuts.

When it comes to the study of crunchiness, no one is more renowned than Charles Spence, who heads up the Crossmodal Research Laboratory at the University of Oxford. In 2004, Spence and fellow experimental psychologist Massimiliano Zampini published a now-famous study called "The Role of Auditory Cues in Modulating the Perceived Crispness and Staleness of Potato Chips" in the *Journal of Sensory Studies.* The researchers used Pringles chips because they are close to perfectly uniform — both within the chip and from chip to chip. They're molded from an alarmingly pliable kind of potato paste. Participants in the study sat in a soundproof booth wearing headphones and were positioned in front of a microphone as they bit into the Pringles chips. The sound of the crunch that came from the chip was routed through the microphone and then fed back through the headphones. Some of the crunch sounds were manipulated — made louder or tweaked on certain frequencies so the participants heard a kind of megacrunch in their headphones. Participants operated foot pedals that controlled a computer-based visual analog scale, which they used to rate the crispness and freshness of the chips they crunched into and heard. Spence and Zampini found that "the potato chips were perceived as being both crisper and fresher when either the overall sound level was increased or when the high-frequency sounds (in the range of two to twenty kilohertz) were selectively amplified." In other words, the louder or sharper the crunch, the fresher the chip seemed. Crispness and freshness is actually in the ear of the beholder.

Since his crunchiness study, Spence has kept at it. In a demonstration that was probably more fun than scientific, he paired up with Heston Blumenthal, the chef at the Fat Duck in England, one of the world's highest-rated restaurants. Together they concocted an experiment meant to measure the way sound influences the perception of taste. It involved Blumenthal's bacon-and-egg ice cream. (He serves it with fried bread — the crunchiness is meant to trigger the experience of crunchy bacon.) At a conference on art and the senses, Spence and Blumenthal had participants taste the ice cream dish and rate its "egginess" and "bacony-ness" while they listened to one of two different sounds. One sound was bacon sizzling, and when it played, people rated the bacon flavor higher than the egg flavor; the other sound was chickens clucking, and when that one was played, participants rated the egg flavor higher. Even though the participants must have known they were being manipulated by sound, they couldn't help themselves.

Beyond providing information and cuing emotions, sound and music can also help you decide how to act. Picture yourself on a morning commute. You step onto a train or bus. There are two single, empty seats, each on the aisle. Both are an equal distance from the exit. Both are the same color and shape. There's a person beside each empty seat, and the two look a lot alike: they are dressed similarly and have the same basic hairstyle, the same complexion. They're both reading newspapers.

Which seat do you choose? What if one of the two nearly identical passengers lets out the slightest cough? That small little cough sets off a whole set of questions in your mind. *Is he sick? Is it contagious? Is he a smoker? Are you going to smell the smoke on his clothes for the whole commute?* Better play it safe and sit next to the person who didn't cough, right? That tiny, involuntary sound conveys enough information to influence not only your feelings about this ordinary-looking person you've never met or spoken to but also your actions.

This power of sound to influence your behavior becomes especially potent when it comes to decisions you make about spending money or time in a place. In a widely cited 1999 study titled "Play That One Again: The Effect of Music Tempo on Consumer Behaviour in a Restaurant," Clare Caldwell and Sally A. Hibbert, researchers at the University of Strathclyde, Scotland, found that diners spent 13.56 minutes longer in a restaurant when they were listening to slower-tempo music than when they were listening to higher-tempo music. They also found that customers spent "significantly more" on food and drink when slower music was playing. Caldwell and Hibbert were building on previous studies that focused on the time and money spent in malls, retail outlets, and cafeterias. In one such study in the 1980s, acclaimed marketing professor and researcher Ronald E. Millman found that supermarket sales went up 38 percent when the store played slow rather than fast music.

A big part of the reason why has to do with arousal — meaning a real, quantifiable response to stimuli. Loud, up-tempo music with unpredictable progressions is really arousing. The Foo Fighters or U2 make you feel something right away. Perry Como? Not so much, unless you count feeling sleepy. (Also, your brain is always looking for sound to fit into patterns it recognizes, and it perks up when familiar pop-music structures are altered. The Foo Fighters, as an example, are masters of hooking listeners with repeated, familiar pop structures and then creatively altering them at precisely the right moment — part of the definition of alternative rock.) When you're more aroused by music, you're less likely to take time to shop, researchers found. You move through the experience faster. Less time shopping means less money spent.

Then there's the sound of the space itself, which can influence your behavior in all sorts of important ways. The sound of a space becomes vital for businesses built on the customers' experience, where the goal isn't just to lure people in but to convince them to stay awhile.

Chipotle, a Mexican-inspired chain restaurant, is the kind of place that's neither cafeteria nor restaurant dining room. Dining rooms sound soft. There's wood involved. Paintings hang on walls. In a cafeteria, you slide a rigid plastic tray across metal rails and request food from someone behind an angled-glass sneezeguard. Your metal chair grates across the linoleum floor. You're there to get fed and get out. Despite its high-volume assembly-line-style service, Chipotle aims to give you a more refined experience than a typical grab-and-go restaurant or cafeteria offers. There, you'll find a sharp visual design palette — white-brick walls, gray or metallic surfaces, and wood.

Most Chipotles in busy cities are in long, narrow, rectangular spaces. You get in line at the back end of the rectangle and move along the wall toward the food station at the front, where you order your burrito or tacos. As you advance in line, you pass diners seated in the middle of the space. The seating arrangement is important for two reasons: It makes the place feel less like a fast-food joint and more like a restaurant, since people are hanging around to enjoy a meal, and it lets you see and smell the food you're about to order, so you start anticipating it. You also hear upbeat music playing across those central seats. "That's exactly why we had to get it right," says Mick McConnell, Chipotle's director of architecture and design.

For a while, though, that music was a problem.

Speakers were mounted in a wooden box directly across from the entrance and queue. The music from them bounced off the opposing wall and created a perfect storm of noise — the kind you might hear in a grade-school gymnasium or an indoor basketball arena, for example. The sound was bright, loud, and brittle — unbearable for people sitting in those important middle seats. The music in other parts of the room was just fine, and customers tended to avoid the seats in the central area if there were tables available elsewhere. So McConnell and his team, including sound engineers, came up with a solution: a perforated metal feature backed by sound-dampening material on the wall opposite the speaker box. It looked on the surface like a simple design flourish. But it was actually a sonic solution. The sound was deadened, and people returned to Chipotle's most important seating area.

My colleagues at Man Made Music and I recently faced a similar challenge. We were asked to create a soundtrack for the foyer of an AT&T store. Not just any store: its flagship innovation store on Michigan Avenue in Chicago. This wasn't a place where people popped in to pick up extra batteries or adapters. It was a showroom for lifestyle technologies that promised to help customers get fit and get healthy, let them express their Cubs and Bills fandom through phone accessories, and offered exclusive phone cases by local designers like Orla Kiely and James Marshall. (Mr. Marshall, a Chicago street artist better known as Dalek, created the well-known Space Monkey cartoon characters.) Also on display was a next-generation, web-based home-security system rigged up in a mock home. There was a connected car of the future on display in another section of the store. Video screens lining the walls everywhere flashed landscapes of Chicago, people walking by, even a cat in a living room — it was all personal and people-centered, not tech-focused, and our sound had to lead it all. So we decided to aim for a sonic welcome experience. It had to be something that would, first and foremost, stand out in a way that stopped people in their tracks and instantly changed their moods at the front door. It couldn't be just music. We had an in-house creative brainstorm, and our team agreed on the idea of developing something like a sonic Rube Goldberg machine, where one sound triggered emotion that triggered action that led to more emotions and actions, with just enough white space (silence) between sonic events to leave people anticipating what might be next. It was that surprise-

and-delight factor. We came up with the idea of two intersecting soundtracks of different lengths: a long ambient track to musically represent the concept of possibility, and another one full of whimsical sonic spikes — fireworks, a whirring machine, a champagne-cork pop, and little reminders of the keynotes in AT&T's anthem (more on the trials and tribulations of creating that anthem later) — performed with a sense of humor. The reason for two intersecting soundtracks of different lengths was so that no one would have exactly the same experience twice. There was a bit of randomness, which made the sounds more interesting for the customers and spared the employees from having to listen to the same repetitive sounds all day. There were all sorts of sonic points of interest to grab on to. We knew that once we got people hooked, we could begin to tell them a story right there in the store's foyer. Within three seconds, they'd be primed for a new type of personal-retail experience, even before they saw a single product. The store quickly became celebrated for its design innovation, winning numerous industry awards, but even more important, it became a beacon project for the company in terms of putting the focus on human outcomes and customer benefits rather than on the technology.

Too often the sonic component of spaces like AT&T's innovation store or Chipotle restaurants is an afterthought, something companies ignore at the beginning then try to correct at the end. Sound is the soft part of the budget, they think. They're wrong. Actually, sound is often the most highly leveraged and efficient element of design, if you consider its cost relative to its impact. Think about the price of other elements like lighting, signage, materials, and construction. Sound is a bargain. And it's the only element you can change literally in an instant to dramatically modify an experience — be it seasonally, or to reflect a time of day or special circumstance. Many businesses strive only to make the sound of the space and the music bland enough so they don't offend anyone; the result is that no one feels *anything* from the sound or music. It doesn't influence anyone's thinking or actions. The space might as well be silent, and businesses ignore massive opportunities.

"Most supermarkets are about as entertaining as a funeral," says Dr. John L. Stanton, professor of food marketing at Saint Joseph's University. "You're kind of lost on an island just trying to pick up things." You're probably all too familiar with the bland easy-listening soundtrack most supermarkets play by default. Some grocers have started to experiment with playing certain music during times of the day, Stanton says: swing or big-band music in the morning, when the clientele is presumed to be older; upbeat rock or pop in the evenings, when young people are stopping by to grab dinner. It's a good first step.

Then there are supermarket chains like Publix in Florida. If you've found yourself there — or at a select Safeway, Piggly Wiggly, or Kroger — you might have encountered a fantastic potential use of sound, albeit one that's unrealized. You'll hear it in the produce aisle: a curious rumble coming from the direction of the spinach and cilantro. It's the same kind of sound that makes you spin around to spot sizzling fajitas or Ford's Bullitt Mustang.

At Publix, "Where Shopping Is a Pleasure" (you can probably sing the slogan if you grew up in Publix country), the rumble is followed by a sudden crack and a boom. Then comes the sound of rain. You turn your head and see what looks like a storm raining down on the carrots and Bibb lettuce. There's a rain forest right there in aisle one! You might have zero idea whether this does anything positive for the food as it sits on a shelf. But it *sounds* fresh. And this isn't some random stockperson blasting your salad with a garden hose. This is *nature*. I mean, it's not really nature. But it sounds like nature. And you can't help but think about those carrots and that lettuce, now miles from any farm, still thriving, soaking up a cool drink from the clouds above.

Jack Corrigan won patents for the automatic vegetable mister in the 1970s and founded the Corrigan Corporation of America back then, after trying out the system in his own produce stores, Carrot Top. The company now provides equipment to about half of all U.S. grocery stores and almost all of those in Canada, says Mike Corrigan, Jack's son and president and CEO of the company. The systems were wildly popular, he says, but if a complaint ever did come in, it was usually from someone who didn't know the mister was about to come on and ended up getting soaked while reaching for romaine. Initially, to address this, Mike and his team created a recorded spoken message, warning people that the mister was about to turn on. But he saw an opportunity to do more with sound. So he developed the thunder track you hear today. His Que audio system, an add-on to the Corrigan mister, was also able to play songs, such as "Singin' in the Rain," before the system doused the produce. When it did, Mike observed something powerful.

"If I was in a store, and the sound played, I noticed customers who were shopping in that department looking at each other and smiling," Mike says. "If you get people in a good mood, they'll buy more." The systems go for as much as five hundred dollars with all of the motion detectors and full stereo sound. But they start at around two hundred and fifty.

Mike created a tool. It's up to stores to use it as part of a strategy. The Que system hasn't been quite the boon to the business Mike hoped it would be, he says. "We've been doing it for fifteen years. We got to the point where it's a gigabyte of sound on an SD card. They can download their own sounds on there. And nobody takes advantage of it." Mike sees a tremendous opportunity in sound that's being missed by many of his customers — only about 5 percent of his mist-machine customers buy the sound add-on. "We kind of walk them down the aisle to a point," he says. "But you can't make them go the next step and realize what a marketing tool this is for your specific store." It's one thing to create a system for delivering sound. It's another to convey the value of using the sound strategically to retailers used to playing bland overhead music. It's the same challenge Jim Reekes faced at Apple. Unfortunately, Mike Corrigan can't sneak in in the middle of the night and install more sounds in all of his grocery-store clients' systems. Once again, smart people are missing massive opportunities.

"Supermarkets are an industry characterized by everybody wanting to be first at being second," says John Stanton, the food marketing professor. Take Publix. A rep for the store contacted for this book couldn't (or wouldn't) explain the store's motivations for rigging up the rain-and-thunder sounds. Neither could the Food Marketing Institute, whose representative suggested the sound served merely as a warning that the mister was about to come on, little more. Thunder in the produce aisle is a classic example of a boom moment unrealized. Sound for sound's sake is often a missed opportunity. People are often satisfied with a sound's being neat or catchy, but they don't realize just how meaningful sound could be if it was applied to drive experiences everywhere. Take the grocery-store opportunity even further. What if, instead of all of that soft rock on the overhead sound system, supermarkets played more nature sounds perfectly paired with the special of the day? What if the big game was broadcast in the beer or chips-and-dip aisles? Or, sticking to the music alone, what if a store broadcast a mix of world music subtly underscoring the experience in the ethnic-foods section? The boom moment could be leveraged again and again all over the store with relatively little, if any, additional cost or effort.

What if the sound at Publix wasn't aimed at the food so much as at the experience? Then Publix might feel a lot more like Jungle Jim's International Market, near Cincinnati, Ohio.

Started as a produce stand in a parking lot in 1971 by James O. Bonaminio in Hamilton, Ohio, Jim's is the paragon of a family business. James goes by the nickname "Jungle" among store employees, in part to distinguish him from his son Jimmy, the business's director of creative services. Jim's has grown far bigger than most mom-and-pop places. There are two locations now, in Fairfield and Eastgate, but the original store in Fairfield, which opened in 1974, now spans six acres. Produce alone takes up an acre.

From the moment you arrive, before you set foot in Jungle Jim's, you hear the noise of an outdoor jungle scene, including the sounds of animals and distant drums. You hear the splash of fountains — elephant sculptures spray water from their trunks in a crystal-blue pool. Inside, you might hear "The Lion Sleeps Tonight" by the Tokens on the overhead system. But there's all different music playing throughout the store, from the sixties, seventies, and eighties stations on satellite radio. The music is layered in a way that makes it crescendo or decrescendo in different sections, creating a dynamic sonic landscape, calling for your attention at times while leaving you alone at others. Even when you're head-down digging for the perfect pepper or cantaloupe, you still hear Jim's personality.

In the produce section, you're greeted by a human-size, talking, Yankee-accented animatronic ear of corn and a companion stick of butter who crack jokes. In the cereal section, robot versions of the Lucky Charms leprechaun, the Honey Nut Cheerios bee, and the Trix rabbit form a three-piece combo — the General Mills Cereal Bowl Band — and they serenade shoppers with pop rock. You also hear Elvis, the pompadoured animatronic lion from Chuck E. Cheese, now repurposed as a mascot in the candy section. And over flute music, robo–Robin Hood welcomes guests to the English foods section from his perch in a sculpted scene of Sherwood Forest.

Jim's also carves out a few spaces that are meant to be perceived as quiet: the wine cellar and walk-in humidor are two of the quietest spots in the store. The sound there is the calming whoosh of air keeping things cool.

(Though sound isn't the focus in the bathrooms, they're an experience too — they look like portable toilets from the outside, but they're actually portals into full facilities; the effect can be jarring when you see groups of people exiting together from what appears to be a single-person porta-potty. In 2007, the bathrooms were voted the number-one restrooms in America by Cintas, makers of commercial bathroom products.)

At the end of the day, Jim's is a supermarket, and there are typical supermarket sounds — clanking carts, boxes being ripped open, cellophane bags rustling, the suction of refrigerator doors opening, glass clinking. But you might also hear clapping from attendees in a cooking class in a session upstairs or the amplified voice of a tour guide leading a group through the store; a playful page over the PA system for a manager referred to as "Frankiepoo"; or the dull thud of a wooden mallet being used to speed the demise of a fish that was just netted from Jim's live tank. You'll get the occasional chirp of a bird that's found its way inside the massive place mingling with the sounds of three robotic blackbirds. And you'll always hear people speaking languages other than English, says Annie Burkhart, the store tours manager at Jungle Jim's. "We are part of a multicultural world, and real food is one of the most enjoyable and memorable parts of it," she says.

Fifty thousand people a week shop at Jim's. It reportedly pulls in almost ninety million dollars in annual revenue. And no one is worried about Walmart stealing Jungle Jim's customers, even though there are *eight* of them within a ten-mile radius. The closest is just a few miles down the road.

Jim's has grown big enough to have its own zip code and buy in bulk the way massive chains do, but size and price aren't the main selling points here. The entertainment at Jim's doesn't have nearly as much to do with food as with the experience of grocery-getting. And that's the point: Sound — and visuals, aromas, and more — transform a chore into a trip to a theme park.

4

The Principles of Sonic Branding

S ONIC BRANDING IS the label people in my profession put on the process of achieving tangible effects with strategic, tactical holistic uses of sound. While sonic branding might be a relatively new term, smart intuitive creators of sound have experimented and documented best practices for decades. But even in this age, when people are obsessed with codifying all kinds of things into six-, seven-, or twelve-step programs, many still fumble their way through sound. They bolt it on without thinking thoroughly about the real experience it sets off (or ruins).

It's easy to believe you already know how sound works — after all, just about any human can tell when he or she is hearing a meaningful piece of music or a catchy jingle. Think about pop music. An upbeat Pharrell Williams or Katy Perry song may seem incredibly simple. But making a pop song is a challenging and nuanced process. A lot goes into the creation of an irresistible pop hook. And when you need that sound to do even more — to align with a brand or convey a specific message — now you're talking about a serious craft.

This chapter outlines the basic principles of sonic branding. They can be applied universally — they're just as effective for brand managers, marketers, and advertisers as they are for business owners, architects, party planners, and anyone who does anything with sound.

USE SOUND TO CREATE EXPERIENCES

The idea behind branding is that companies, makers of things or services, can tell stories or position their products as useful components in those stories. The most important question to ask about a sound is whether it makes you *feel* anything. Does it call to mind a specific memory or story? If not, that sound probably isn't fulfilling a precise, strategic function.

Some brands use sound and music to connect with cultural experiences and become part of the nervous system of a community. Look at how Doc Martens became the de facto footwear of punk rock, how Chuck Taylors became the shoe of indie rock, how Vans became the sneakers of skateboarding and surfing culture, or how Adidas became the go-to kicks in the golden age of hip-hop.

Others fail spectacularly at this. They make the mistake of confusing music their audience likes with music and sounds that create a more meaningful emotional connection with their story.

Aniruddh Patel, associate professor of psychology at Tufts University, has studied the curiously complex way the experience of sound plays out in film. Several of his experiments with his colleagues involved taking one film clip and substituting the score of another — something we've all seen done on YouTube, with hilarious results. But when Patel and his colleagues measured changes in viewers' perceptions, they found the impact went far deeper than they expected. "The obvious thing that we might all think is that the emotions change," Patel told an audience in 2013 during a presentation by the World Science Festival and the New York Philharmonic. But "the music, it's not just intensifying the mood. The music can create expectations about events that might be coming up next. It can lead to associations that aren't necessarily on screen. It can [help you] tell if the story is resolved or not. It can change the way you interpret characters in terms of their relationships. . . . It can really influence our sense of being inside the film."

Even when it isn't connected to a visual story, music can create a complex experience that doesn't always match the message on the surface.

Elton John got this one right when he sang "Sad songs say so much." While sadness might be the subject of lyrics, or the song itself might be considered sad, the experience of listening to those songs can elicit positive emotions, according to a June 2013 study by Japanese researchers titled "Sad Music Induces Pleasant Emotion." The researchers played music excerpts and asked forty-four participants to rate the emotions they felt. "The results revealed that the sad music was perceived to be more tragic, whereas the actual experiences of the participants listening to the sad music induced them to feel more romantic, more blithe, and less tragic emotions than they actually perceived with respect to the same music," they found.

The content is one thing, but the experience is another, and it's much more powerful. This principle plays out in all kinds of scenarios.

In 2010, for example, executives at Harman International, the maker of JBL speakers and other brands in the multibillion-dollar consumer and professional electronics business, were freaking out. A bunch of big stories in the course of about a year claimed that modern technology consumers cared less and less about quality, particularly when it came to music. In one of the biggest such stories, published in 2009, *Wired* writer Robert Capps called it "the Good Enough Revolution." His story gave the Flip Cam by Pure Digital Technologies, a simple yet powerful video recorder, as an example of technology that didn't offer superior quality but was successful mostly because it was easy to use. (Flip would be bought by Cisco Systems before being discontinued in 2011.) Other examples of "good-enough" tech included drones instead of piloted jets in Middle East wars and do-it-yourself legal-document kits instead of actual lawyers. But a linchpin of the story involved digital music files on iPods — Capps argued that consumers had become so accustomed to compressed MP3 files that quality was now a slave to portability.

The *Wired* story cited a study by Stanford University music professor Jonathan Berger, who spent six years asking new students in his class to listen to the same music on different digital formats. Berger concluded that students actually *preferred* the lower-quality digital music files over the higher-fidelity ones. "They've grown accustomed to what Berger calls the percussive sizzle — aka distortion — found in compressed music. To them that's what music is supposed to sound like," Capps wrote.

This deeply bothered Harman execs. The company had staked its business on quality — its marketing tagline was "Where Sound Matters." What if sound didn't matter? What if they were wasting millions on tools such as fake ears to measure headphones' fit, near-silent anechoic chambers, and rooms with pneumatic-speaker shufflers that made sure speakers were arranged perfectly for listeners? Harman even built a dummy with binaural hearing (named Sidney, after Harman's founder) to test sound in cars. Could the company's entire business model be flawed? they wondered.

At the same time, the audio electronics market was experiencing a minor disruption from a year-old headphones upstart called Beats by Dre. Audiophiles quickly dismissed Beats as being too heavy on bass and muddy on other acoustic measures. Nevertheless, they were selling like hotcakes. Harman blamed this on marketing. Beats did have celebrity endorsements — gangster-rap pioneer Dr. Dre and Interscope Records chief Jimmy Iovine are cofounders, and pop-music royalty will.i.am and Lady Gaga are investors and endorsers of various models — plus, Beats made sure lots of celebs wore its headphones on red carpets and in videos (will.i.am actually wore a pair during a video shoot for the project I produced with him to refresh the venerable entertainment news program *Entertainment Tonight*). "They present them as a lifestyle fashion item, like a pair of jeans or a pair of Nike shoes," said Dr. Sean Olive, Harman's director of acoustic research. "The headphone becomes associated with this cool — it looks cool if I wear it around my neck. It's associated with a successful, hip-hop, urban lifestyle. That is a powerful message to the people who buy it."

But Harman and Olive were missing the big picture. They were focused on the sound, not the experience. Olive zeroed in on the claim, made by Berger, Capps, and others, that kids actually *preferred* bad sound. "I think our human ears are fickle," Berger told the *New York Times* in 2010. "What's considered good or bad sound changes over time. Abnormality can become a feature." The assumption, particularly with Beats, was that they satisfied young people's need to hear a bunch of bass over everything else. Some had gone as far as to suggest that booming bass is the modern sound of youthful rebellion ("Bass has *always* been the quickest way to piss off your parents and dazzle your eardrums," wrote Slate.com's Jesse Dorris in a 2013 story about the success of the headphones.)

Olive saw an opportunity to attack the science and, maybe indirectly, defend his job. He set up a double-blind study to prove that quality still mattered. Through a rigorously tested and verified process, he cloned the equalization and overall sound of several headphone types, including Beats by Dre, and ran each of those cloned sounds through a single generic over-the-ear headphone worn by test subjects. The idea was to control for form and feel and get right to the type of sound that both college students and trained Harman listeners liked best. The preferred headphone sound was neutral, spread across the audio spectrum, Olive found: "There was no evidence that these kids preferred headphones with boomy, bass-heavy sound." Using that and a handful of other tests, Olive and his colleagues proved the opposite of what *Wired* had argued and Stanford's Berger had found in his study. They ranked Beats last or next to last on a variety of measures associated with the concept of "good." In a scientific setting, kids didn't prefer bad sound. Also, to hear Olive tell it, Beats kinda sucked.

So Harman stayed its course. Its JBL brand rolled out a new high-end line of headphones, Synchros, similar in price and meant to compete with Beats by Dre's studio model. The big feature of the JBL headphones was LiveStage, a technology meant to make the sound heard through its headphones seem like it was coming from all around the listener, not just in his head. Almost anyone with ears who listens to music with a pair of Synchros headphones can hear a remarkable difference in quality between these and headphones without LiveStage. To announce the new product, Harman had an open-bar event for a couple dozen members of the media and Harman's own sales team in a basement meeting room at the hip Ace Hotel in Manhattan. The headphones were set up on folding chairs in front of a stage. The idea was that guests would later put them on and listen to a performance by emerging singer-songwriter Trixie Whitley. Instead of listening to sound coming from amps or instruments on a stage, event-goers would all hear the performance through their Synchros headphones, a sort of silent concert. Anyone who tried to put on the headphones or test them out on his or her own iPhone *before* the presentation (which included a thirty-minute PowerPoint talk by Dr. Olive) was scolded by a company rep.

Despite all their research, Harman and *Wired* and Stanford's Jonathan Berger all made the same mistake. They presumed that people were making choices about sound based on the sound quality. They weren't, at least not primarily. A representative group of college students might have demonstrated in a lab that they preferred quality sound. But they didn't live in a lab. They were listening on the go, multitasking, using sound as a component in all kinds of experiences. The sound of Beats fit into the idea of how they lived their lives, their lifestyles.

Additionally, users weren't choosing iPhones and iPods because they preferred the sound of the MP3 format. They wanted beautiful-looking devices with dead-simple user interfaces that could fit in a pocket and transform almost any mundane experience into a musical one. As a bonus, these stark, rectangular gizmos made statements about users' appreciation of design. These weren't MP3 people; they were Apple people. Saying that people who listened to these portable music devices preferred low-fidelity sound was like saying kids (or adults) who drop everything and sprint for the ice cream truck prefer low-quality treats. Hold a blind taste test between Mister Softee and Ben and Jerry's, and Chunky Monkey wins every time. It's not about the ice cream, and, for Beats, it's not about the sound—it's about the experience. The Beats company aligned itself with a community built on a certain sound—not just faceless low-frequency bass, but icon Dr. Dre's sound, the godfather of gangster rap whose touch as a producer influenced several generations of hip-hop. Distinctive-sounding (and -looking) headphones give people a way to put themselves in that story and ally with a community. Beats by Dre headphones don't boast the most acoustically precise listening experience, as Dr. Olive's double-blind tests confirmed. They offer the right kind of sound for the right group of people at the right time. Bass-y, low-end sound is what you expect in a club with stars of hip-hop and pop. You put the headphones on because they make you who you want to be.

Harman and their counterparts in consumer audio electronics have a deep understanding of the *science* of sound, and a healthy segment of consumers will always look for quality. Beats gets the *power* of sound to drive an experience. By late 2013, Beats had launched ten models of headphones, plus three different speaker systems, a branded HTC smartphone, a laptop and monitor with HP, and car-audio systems for Chrysler, Dodge, and Fiat. They all bore the iconic lowercase-*b* logo. And they all offered a sound that was boomier than similar devices', because that's part of the Beats story.

At the end of January 2014, the company launched Beats Music, a streaming-music service most assumed would compete with Spotify, Rdio, and others. But like the headphones, Beats Music was just a piece of the puzzle for a full-fledged lifestyle brand built on the idea of creating a music-driven experience.

"What we have right now, fortunately, is we have a very trusted name in music with Beats," cofounder Jimmy Iovine said in a conference call prior to launching the service. "We feel that it was step one to developing a complete musical thought."

Harman, which has been around since 1953 and produces a broad array of professional and lifestyle audio products for homes, cars, studios, clubs, and more, expected to bring in as much as $4.25 billion in 2013. In late September 2013, Beats, which has been around only since 2008, announced a five-hundred-million-dollar investment from the Carlyle Group and was expected to pull in $1.2 billion in revenue for the year. It owned 64 percent of the market for headphones costing more than one hundred dollars, according to the NPD Group.

When sound is used to create experiences, it gives rise to a tremendous advantage, even for a six-year-old, muddy-sounding headphones company. Then in May 2014, reports surfaced that Apple was planning to acquire Beats. The rumored price? Three point two billion dollars.

MAKE SOUND MATTER: CREATE ANTHEMS, NOT JINGLES

Getting the right attention and keeping it requires a commitment to a meaningful aesthetic experience. This is true whether you are developing a workout or party playlist, a soundtrack for a video game, or a sonic strategy for a multibillion-dollar company. The latest pop hit might make your heart skip a beat, or a novel or loud sound might grab your attention for an instant, but unless you are clear about the experience you are trying to create, people won't remember it and you won't forge a connection. In the words of Maya Angelou, "I've learned that people will forget what you said, people will forget what you did, but people will never forget how you made them feel."

This is, again, where strategy comes in. If you're looking to associate your story with music or sound already out in the world, then you really need to know what your story is — why it's important, how it's different than someone else's story, how you want people to *feel*. Only then will you have something to measure the music against. Then you can ask, "Does this sound like my party [or film or show or store or brand or cause], or someone else's?" Maybe you don't own the music you use, but if the curation *means* something, and you do it with a great deal of integrity and consistency, then you can end up like United Airlines, Target, Apple, or world-famous DJ Tiësto — with a clear musical zeitgeist that surrounds and resonates with your story.

If you're looking to create an original, ownable sonic identity for your story or your brand, and you're not creating an anthem, you're missing a huge opportunity. Think of the way national anthems can inspire people. Many sonic logos, especially in the form of snippets of sound dropped in almost exclusively at the ends of ads, are meaningless. You might remember them or recognize them, but they tell you nothing because you feel nothing. Without an associated long-form anthem, a sonic logo has no emotional memory to trigger. There's no context, no story, and usually no strategy.

It's important to understand the distinction between jingles and anthems. A jingle can be catchy, but only an anthem can carry a complete, emotional story. It packs the sonic themes that can be distilled, adapted, and reinvented into all sorts of musical styles. It's part of the glue that holds together movie franchises like the Harry Potter, Star Wars, James Bond, Lord of the Rings, and Pirates of the Caribbean films. In a sonic-branding process done right, logos and jingles are distilled from anthems. In Beethoven's Fifth, the section that goes *bum-bum-bum-buuuuuuuum* is the motif, which is analogous to the logo — it's part of the anthem. You might not be able to hum the rest of the tune, but you inherently understand that the line belongs to something bigger.

Think about the seven notes of "When You Wish upon a Star." They show up everywhere from the opening sonic logo of Disney films to the cruise-ship horn on the Disney Cruise Line. They remind you of the whole song in its context and, more important, how that song makes you *feel*. You hear those notes and your mind jets back in time to Disney's *Pinocchio* or a trip to a Disney theme park where you posed with oversize furry characters in front of the Magic Kingdom for a picture that still sits on your mom's mantel. If you haven't done the strategic work of truly understanding your story and creating an anthem that's adaptable for all kinds of actual experiences with your brand, your logo probably reminds listeners of . . . your ad. Or, worse, it packs no emotion or memory at all. A logo or a jingle alone doesn't tell a whole brand story. And if that's all you have, you're doomed to fail. At the very least, you'll miss out on a lot of opportunities to help people *feel* your brand.

Consider what a sonic logo sounds like when it isn't attached to this kind of anthem or story. The television network ABC has used the same four notes for much more than a decade. Can you sing them? You can sing the three chimes of NBC, but that's because they've been baked into hundreds if not thousands of anthem and theme contexts and rearranged and reharmonized many times since 1929. Poor CBS tried to duplicate this success but failed, and Fox never even tried.

Once you have a strategy and an anthem, the basis of a sonic-identity system, then you can map the true DNA of that anthem and start using it to heighten people's experiences everywhere your brand goes. Where can sound make a big difference for your customers, your employees, or your partners? Be more like a big movie franchise, and *score the brand experience.* Think about all the places you see McDonald's branding. In how many of those instances do you hear it or feel it? Outside of the last three seconds of its television commercials, the brand is essentially silent. In most of McDonald's touch points (all the places where customers come in contact with the McDonald's brand), you could be hearing almost anything. If McDonald's used sound to its full advantage, people could be hearing and feeling its sonic story in overhead music in the stores, at live events it sponsors, from the toys it gives away, on its website, and more. The corporation could build contests around it, bring in artists to rerecord it, have it greet you at the door if you're the thousandth visitor of the day. But McDonald's doesn't do any of that. It's a huge missed opportunity for one of the smartest marketers in the world.

CURATE YOUR SOUNDTRACK

Soundtracks aren't just for movies. Everyone has his own personal soundtrack, whether it's the playlist streaming through his earbuds or the noises made by the people and objects he encounters in the course of the day. Sonic branders align with those experiences and amplify or steer them with musical moments that call attention to the right truths at the right time. The smartest ones find these opportunities everywhere.

Sound is really the emotional engine for any story. The soundtrack of a film triggers emotions, but research shows that it also does much more. If you change the score of a film, it can dramatically alter people's view of the relationships between its characters and their expectations of where they think the story will go. Music also triggers memories and connections. Hearing a musical theme that was established earlier in the film can suggest a sense of familiarity and poignancy. It's a kind of short-term nostalgia that sound best captures.

Horror-movie makers are the masters of using sound to drive emotion. Which is why, if you feel yourself getting freaked out in a scary film, you should cover your ears. You won't miss what's happening, and you won't get startled. Want a laugh? Watch the terrifying trailers to any of the Paranormal Activity movies with the volume turned down. Try to make it through a silent version of *Halloween* or *Friday the 13th* without falling asleep. The emotional and musical language of horror is actually extremely sophisticated in terms of action, story line, and character development. The soundtrack does a lot of heavy lifting. On a rudimentary action level, there are scenes where the soundtrack helps build tension, often with a moment of silence, which then sets up the scare. Equally powerful is the building of tension leading to a fake-out (no scare), or sometimes even a *fake* fake-out (no scare, quickly followed by a scare). With a fake fake-out, you relax after that tension only to be shocked by an unexpected jolt.

The cat, not the killer, jumps out. Then the killer leaps out from a different dark corner right afterward. John Carpenter, composer of the score for the Halloween franchise, is a master of this.

The right sounds can also relax you. Ambient background music can be a powerful tool for creating a bed of soothing sound. It's one of the secrets of W Hotels. Notice the sound at any W or other boutique hotel the next time you visit. The smart ones play tunes in the restaurant, poolside, and in your room as you first check in. It doesn't really matter if you love the music itself. Think about what you might hear if there was no soundtrack playing as you arrived in your room: the noise from the street below, the clanking of the air-conditioning unit on the building next door, or the conversation of the couple in the adjacent room. Hotels can't always control environmental sounds, but they can at least drown them out.

The next time you notice a sound, think about whether you're supposed to. Is it sticking out because it's too loud? Or did you notice it because it didn't fit the story? Was that on purpose? Most important, do the sounds you hear make you feel anything at all? Asking these questions can help you become more aware of the ways that sounds are manipulating you. Whether or not you welcome that manipulation will depend on the situation. Noticing who's good at this and why is the first step to figuring out how to more effectively curate the sounds you make and the way people hear them.

MAKE SOUND WORK HARDER

Short sounds are all around us all the time. I often refer to them as sonic triggers, because they can be the sparks that lead to physical actions or memories or feelings. You've probably responded to one in the past few minutes. Close your eyes for a minute if you're outdoors or in a crowded environment, and you'll probably hear one again within a few seconds. These triggers are defined by their ability to convey a lot of information in an instant. Memory is a big factor here — what experience, recent or distant, do you associate with that short sound? New moms and dads will tell you they suddenly have heightened reactions to the sounds of babies crying — they hear it everywhere and it almost always startles them. The chirp of certain birds — a whip-poor-will or loon, for example — can call to mind particular times of year at specific places, probably by a lake or a body of fresh water. The sudden symphony of the wings of cicadas or locusts in the middle of the day can conjure memories of hot climates and summer months.

When sound is working at its highest potential, it surprises the ear. Have you ever been to a concert where the band has been playing all night, and then, just as they are nearing the natural peak of a song, it happens: *wham!* One note or beat radically changes the experience in a rapturous way, and you soar off into a completely unexpected musical direction. Chills run up your spine and you feel that rush. That's the band's boom moment. And it's built on surprise. As listeners, we tend to hear music in *moments*. They delight us as they extend the story of a song we think we know, and they break an expected pattern in our brains. Whether you are moved by the unexpected sound of Pete Townshend's guitar windmills, Haydn's Surprise Symphony, a cunningly altered note in a Charlie Parker sax solo, or the distinct growl of the pipes on Steve McQueen's tricked-out, definitely-not-standard-issue Mustang in *Bullitt,* these are all moments where expectations are set and then abruptly broken, resulting in a rush of endorphins and an addictive desire to have it happen again.

Sonic surprise is one of the oldest sonic tricks in music, movies, entertainment, and even advertising. You hear it whenever an actor's voice is intentionally replaced by the voice of someone from the opposite gender or someone younger or older than him or her. Or when an expected animal sound is replaced with a human one. It gets your attention. But getting someone's attention is the easy part. Effective sonic branding often involves creating or facilitating sonic triggers that break expected patterns, get the listener's attention, and then using that attention to call to mind positive experiences with the brand or story.

We already recognize these kinds of snippets in the real world, and we pull information from them all the time. After all, what is a car horn, a referee's whistle, or a shot-clock buzzer? How about the school bell? Think about the satisfying sound a golf ball makes when you sink it or the clean successful swish of a perfect three-pointer in basketball. Each of these causes a thrill while confirming something or telling you what you need to know in the moment. They're almost Pavlovian, except they're initiating something far more complex than physiological reactions. This is bigger than drooling.

It's easy to use sonic triggers to turn a forgettable experience into something memorable and meaningful. In fact, you already do. You yell "Surprise!" to make a birthday party boom. You clap louder and shout "Bravo" when a performer is especially inspirational or above and beyond the expected. I know of a record label's radio marketing team that created their own boom moment to signal success when they were trying to get radio stations to play a new song they were targeting. Every time a major new radio station added that song to its playlist, the head of sales invited the person who made it happen to hit a gong, centrally located in the office. The short sound would cause everyone to cheer and try to figure out who had had a big success. The whole group felt optimistic about the win, which led them to more wins. It was a huge way to build office camaraderie and a culture of supporting people's success.

Sonic triggers can also be deployed as effective functional sounds. They can be a welcome or a reminder, provide vital information, or help you understand where you are. My mom used to ring an actual dinner bell. This bell served two purposes: First, I could hear it anywhere in the neighborhood. When I did, I'd instantly feel disappointed—I had to stop what I was doing and come home. But second, it reminded me that I was hungry, so I'd get excited and run home, hoping for mashed potatoes and meat loaf.

When you think about the short sounds cars make, you probably think about revving engines, horns, or squealing tires. You most likely don't think about the purr of an idling or a near-idling engine — but you'd think about it if it weren't there and you suddenly found yourself in the middle of an intersection staring at the grille of an oncoming electric car that you didn't hear coming.

In 2009, the National Highway Traffic Safety Administration (NHTSA) found that when slowing, stopping, backing up, or leaving a parking space, a hybrid-electric vehicle was two times more likely than a vehicle with an internal-combustion engine to hit a pedestrian. In January of 2013, the agency proposed new rules requiring electric vehicles going less than 18.6 miles per hour — that's the speed at which the sound of an electric car matches the sound of an idling internal-combustion engine — to emit warning signals that walkers, cyclists, joggers, and blind people could hear over typical background noise. The NHTSA said these warnings didn't have to sound like the annoying beeps of reversing commercial vehicles. (A noise likely to find you whether or not you're in the vehicle's path.) Carmakers would get to choose their own signals — rarely does government regulation present such opportunities for brands to have boom moments. Audi found one with its e-tron electric sports car; its engine-rev noise sounds like the light cycle in the Disney movie *Tron*.

These sounds, the NHTSA said, would save 2,800 pedestrians and cyclists from injuries for every model year of electric and hybrid vehicles. The rules were scheduled to go into effect in September 2014.

This is far from the only instance where sounds are perfectly suited to carry special meaning and save lives based on when and where they appear.

A Danish audio software company called AM3D markets its binaural headwear apparatus to firefighters; it allows them to know, in smoke-filled, nearly blind environments, where their team members are. They hear three-dimensional, spatially accurate sound cues. The company also markets the tech to the defense industry. It's used in some A-10 and F-16 fighter aircraft. Pilots receive audio alerts about missiles or enemy fighters through speakers in their helmets, and in addition, the sound tells them, in an instant, what direction threats are coming from (including above, below, or behind them). Reacting to a visual warning takes about a second or so longer than reacting to an audio cue, and in a situation where a pilot has about five seconds to react to an incoming missile or enemy jet, that extra second is potentially lifesaving.

When used in gadgets, short sounds are typically called user-interface sounds. In my business we call them brand-navigation sounds. The term reminds us that they must be both emotional (the *brand* part) and functional (the *navigation*). The sound has to work harder, creating a sense of identity with the brand *and* making the technology more intuitive for the user.

The creators of one of the most popular games on the planet, Call of Duty, made short sounds work harder and become essential in gaming — gamers use the sounds to recognize surroundings and advance through levels. The makers of Angry Birds created a satisfying crunch and squawk to make you crave endless rounds of play, then used those sounds to trigger the same cravings in an endless array of toys and products and media. Product engineers and designers such as Jim Reekes created short sounds for Apple's early Macintosh machines — most notably, Reekes's Zen-like start-up sound, which lives on in the Macs we know and love today. All of these pioneers know a secret: we're entering an era of hard-working short sounds that help guide our experiences with everything.

SILENCE IS POWER

Humans aren't born with *ear*lids. So for most of us, there is no such thing as true silence. Want proof? Put on John Cage's *4′33″*, a piece of conceptual music without a single note. It's meant to make us aware of the sounds constantly playing around us. No matter where you are, no matter how quiet you think that place is, when you sit in it for a while, you'll start to hear this symphony. Probably before Cage gets to his first chorus. (That's a sound-nerd joke.)

I spend a lot of time on this, along with my team at Man Made Music. A vast majority of our clients don't initially get it right, even the smartest of them. If sound or music is good for an experience, then more must be better, right? The fact is that often we spend weeks, months, or even years trying to remove meaningless or pointless sound from stories, shows, spaces, and brands.

Disney, however, is a company that gets sound. And it takes an exemplary approach to creating silence at Disneyland and Disney World. As long as guests are inside the park gates, they're not supposed to feel like they've left the fantasy. Still, not even Disney can control nature or the airspace above its parks. And nothing shatters the illusion of an old-timey Frontierland fantasy faster than a low-flying 747 jumbo jet. To guard against magic-killing noise, "We have to create a perceived quiet," says Disney's principal media designer Joe Herrington. "A forest and bird bed of sound can be perceived as quiet."

That's right. Disney uses sound to fake quiet.

It's a more powerful tool than you might think. Disney's fake quiet also works as a natural barrier between its lands. It would be jarring to hear the sounds of Tomorrowland while you were still strolling down Main Street, U.S.A. So perceived quiet in the parks lets you walk a hundred yards, cleanse your sonic palate, and feel like you're entering a whole new world. How much farther apart would the rides need to be and how much larger would the park itself have to grow to protect the magic if Disney hadn't learned this sonic secret?

As for pure quiet, you probably couldn't handle it. After about thirty minutes in a sound-swallowing space called an anechoic chamber, you might start to feel like you're losing your mind. The Orfield Laboratories in South Minneapolis is considered by Guinness World Records to be the world's quietest place — 99.99 percent of sound is absorbed by foot-thick concrete walls with insulated steel and 3.3-foot-thick fiberglass wedges. Even the floor is insulated this way (you stand on a kind of wire support). No one has lasted more than forty-five minutes in the room without suffering severe discomfort. When you turn the lights off, the flat, disorienting space provides so little feedback in the form of sound waves that you have to sit. And when all of the outside stimuli gets sucked into the walls, the inside stuff gets loud. Your brain, so used to filtering out the constant barrage of vibrations from the world, starts picking up tiny sounds and bringing them to your attention. After just a half hour or so, you start to hear your own heartbeat, your organs squishing, the air moving in and out of your lungs, your joints creaking. "In the anechoic chamber, you become the sound," lab founder Steven Orfield told the UK's *Daily Mail*.

So if we eliminate the myth of silence, what is the opposite of noise? It's something more akin to white space, like in visual or print design. Silence is contextual, not absolute. And our brains are constantly adjusting that context — curating sounds, picking which ones to make conscious and which ones to file away.

There's another way Disney gets this right. On the Tower of Terror ride at Disney's Hollywood Studios in Florida or at Disney California Adventure, the scariest part is when you shoot to the top of a rickety old elevator shaft and dangle there in your seat for a few seconds. You hear wires and cables start to crackle and snap. You know you're about to plunge.

But in Tokyo, Disney wanted to make the experience even scarier. So they pulled out the sound at the climax of a key scene. Herrington and his team had the idea to use nearly absolute silence to heighten the tension. "The idea wasn't received well at all by management early on," he says. They thought it would take away from the experience. But the sound team convinced everyone to let them try it. You shoot up the elevator shaft just like the rides in Orlando and California. Then the door of your elevator car opens, and you stare off into space, the Twilight Zone, as it were. And the sound just echoes off into the nothingness. Gone are the snaps of wires and splintering metal of cables. In Florida or California, if you ride the ride twice, you know exactly when you're about to drop. Even if you don't consciously count out the seconds, you have a sense. The sound prepares you for action, the way a movie score does. But pull that sound out, and the suspense is multiplied exponentially.

"It is absolutely terrifying, because you know you're about to fall but you don't know when, and it's dead quiet, and then you fall out of that silence. People are terrified. It was a huge hit," Joe says. "Silence is a powerful tool."

In fact, Joe and his team overshot the mark. According to Japanese park executives, the Japanese Tower of Terror profile was too scary. "It'd make you wet your pants," Joe recalls them saying. A year later, when they found that more college-age guests were coming to the park, the Japanese managers asked for three new profiles to give them something special and memorable. "They said, 'We really want to scare them.' It really made an incredible difference," Joe says.

It's important to know just when to take the sound out. Give it a shot and see what happens. If you're working on a product or presentation that involves sound, try taking out individual sound elements and listening to how it changes the overall effect. Turn down the sound at key moments when watching your favorite movie and see how the experience changes. Put on some sound-canceling headphones at home or at work and pay attention to what sounds are missing.

Sound anchors us in our world, gives a sense of what to expect, and completes our picture of ourselves in space. When that's gone, we can be left without any context and that can be terrifying. That can be useful if terror's what you're after. But pulling the sound out can also provide a break, a respite from a barrage of sound that allows our brains to reset before we crave sound once again. When adding sound to an experience, always be sure to pull it out one final time before you're finished. If you don't miss the sound, it probably shouldn't have been there in the first place.

DUMP THE SONIC TRASH

Just as sound is one of the most powerful tools we have to tell a story, the wrong sound is one of the most powerful ways to kill one. Simply put, sonic trash is any sound that diminishes your experience because it's wrong to you. As the multiple Grammy- and Oscar-winning composer Hans Zimmer puts it: "Get rid of the shitty sound. Life's too short."

In January 2010, Frito-Lay debuted a 100 percent biodegradable bag for its SunChips brand. The bag was designed to cut down on landfill waste, but it completely polluted the sonic landscape of customers and anyone within earshot. A Facebook group called Sorry But I Can't Hear You Over This SunChips Bag sprang up and gathered more than forty-four thousand fans. In a report about the bag, an enterprising television reporter for CBS found that, when shaken, the bag registered one hundred decibels, louder than a lawn mower (ninety decibels), a motorcycle (ninety-five decibels) or a subway (ninety-four decibels) — the reporter even shook the bag on a subway platform, and it cut through really loud sounds there. SunChips sales dropped every month, in year-on-year measurements, from the moment the bag debuted. Frito-Lay tried to add an adhesive to the material to cut down the sound. But ten months after announcing the bag, Frito-Lay said it was scrapping the crinkly nightmare. At least we know all of those bags broke down quickly in landfills.

The Oxford psychologist Charles Spence famously discovered how the crunchiness of chips and crinkly-ness of packaging influences perceptions of flavor and freshness, but SunChips took it several steps too far. It's a reminder that sound is never neutral. It always tells a story, and sometimes it's not the story you intend. You ignore it at your peril.

The noisy SunChips bag is an example of what I call sonic trash. It's a complete disregard for sound in storytelling. And in the case of the noisy bag, the wrong story was louder than the one Frito-Lay set out to tell. Other sonic trash can involve sound inserted in the wrong place or sound inserted solely for the sake of filling space when what's really called for is silence. It's Nissan's weird digital doodle at the end of its ads that means precisely nothing and doesn't make humans feel anything but advertising. It's the aggressive score in the otherwise stunning 2013 movie *Gravity*, about a chaotic accident in space. The film is painstakingly accurate about the way things work in orbit, including the fact that you can't hear explosions or metal shredding or glass shattering. One of the effects of losing sound in a situation where people have come to expect it is that they look for visual answers to what's happening (next time you're at an ATM that doesn't beep, notice how much you lean in and pay attention to the screen). Instead of letting that disconcerting silence drive really violent scenes in *Gravity*, the filmmakers stuff the vacuum with strings and music meant to convey the emotions of Sandra Bullock's character. Scoring to her emotions might make sense in a regular film, but this is not a regular film. Just as you start to wrap your head around the physics of a pivotal scenes, the score rudely insists you pay attention to how it all makes Sandra Bullock feel.

We've all been yanked out of a story by a misplaced film sound or song. Think of Hammer rapping "Addams Groove" over the 1991 remake of *The Addams Family*; P. Diddy rhyming over Jimmy Page's "Kashmir" riff on "Come with Me" for 1998's *Godzilla* remake; Limp Bizkit rap-rocking "Take a Look Around" for 2000's *Mission: Impossible II*. These films shoehorn in pop icons with their own stories, which don't align with the stories the filmmakers are trying to tell.

You've also heard what it sounds like when marketers try to get away with a lie. In 1987, Nike and its ad firm Wieden+Kennedy featured the Beatles' "Revolution" in a sneaker ad. There might have been a time when Nike was an upstart rebel company, but that time was long gone by 1987. They paid $500,000 to license the song, but hard-core Beatles fans and the band's remaining members themselves were incensed. Through their record company, Apple, the surviving Beatles sued the shoemaker for $15 million. George Harrison said in a statement:

"Every Beatles song ever recorded is going to be advertising women's underwear and sausages. We've got to put a stop to it in order to set a precedent. Otherwise it's going to be a free-for-all."

The band and the brand later settled out of court — the terms were sealed. And Nike eventually stopped running the ads.

In plenty of other cases, brands latched onto a chorus or a hook of a song without considering the whole story it told. Many have gotten away with it. Their spots didn't offend anyone, even if they faded away without leaving a mark. Now, though, in an age where we're all more skeptical than ever and well aware of marketing, such misuse of sound can become infamous as sonic trash, as was the case with the use of a song for one particular brand campaign, which readers of the online magazine *Slate* named the greatest misuse of music in an ad.

Royal Caribbean Cruise Lines sought to highlight the more adventurous side of its family-friendly fun cruises in 2010. So the company, along with its ad agency Arnold Worldwide, used "Lust for Life," a song originally written by Iggy Pop with David Bowie. "We were using a portion of the song that musically and lyrically fit with what we were doing," Arnold's managing partners and group creative director Jay Williams told the *New York Times*. The goal was to attract more young people to the cruises. "The energy, enthusiasm and raw feel was right," Williams said. But if you recognize the song (it's Iggy's biggest hit, and was actually first released in 1977), you might know it as the opener of *Trainspotting*, a film about heroin-addicted Scots. If you dig deeper, you'll discover that the song's lyrics reference William S. Burroughs's gender-bending liquor-and-drugs-peddling stripper Johnny Yen. (His name's in the cruise-ship ads.) But to the best of anyone's knowledge, Johnny's never been the featured entertainer on the lido deck. And it's a safe bet Iggy Pop won't be doing the cruise circuit anytime soon. Bottom line: The music didn't match the story. And to suggest that a Royal Caribbean Cruise is like vacation heroin is, well, a lie. To be fair, Royal Caribbean's profits did surpass all expectations in 2010, but it also had just invested in shiny new ships.

Then there was Wrangler's use of Creedence Clearwater Revival's "Fortunate Son" in its campaign for jeans. The ad uses the first half of the opening verse, about folks being born to wave the flag. But gone is the second half: "And when the band plays 'Hail to the Chief,' / Ooh, they point the cannon at you, Lord." So a song protesting sending the poor off to slaughter became a patriotic celebration of denim.

Creedence singer John Fogerty doesn't own the rights to his music and didn't approve the ad. Explaining the intent of his lyrics in 2002, he told the *New York Times*, "I was protesting the fact that it seemed like the privileged children of the wealthy didn't have to serve in the Army. I don't get what the song has to do with pants." Craig Errington, director for advertising and special events for Wrangler, told the *Times* the song was "written and produced more as an anti-privilege anthem, as an ode to the common man. We sell millions and millions of jeans to those kinds of people and always have." So why lose the second part of the verse? (*Slate* readers also voted this one among the greatest misuses of music in ads.)

The point is that the right song can help drive home a true story. But the wrong song can make it fall apart. You'll tune out at best. At worst, you'll get angry.

Bad sound, not just bad music, can have this effect too. The voice of villain Bane in Christopher Nolan's 2012 film *The Dark Knight Rises* became a problem when audiences watching early film footage couldn't understand what the masked madman was saying. Instead of driving the experience, the sound demanded too much attention from viewers, who just wanted to kick back and feel the story. Bane's warble got in the way. The voice was cleaned up for the final film release, but not before parody videos depicting an inaudible Bane garnered hundreds of thousands of views on YouTube. Bad accents and hammy dialects can ruin a story. Kevin Costner does not sound like Robin Hood, and he doesn't fit the story we know and love. To understand how an ill-considered voice can torpedo even the most established epic tale, I offer these three words: Jar Jar Binks.

Sonic trash can remind you you're watching something fake. In the 1953 cowboy movie *The Charge at Feather River*, Ralph Brooks, playing the character Private Wilhelm, gets struck with an arrow while riding on horseback. He lets out a scream you've surely heard, whether or not you've actually seen the film. The Wilhelm scream, as it's become known, was dubbed in by sound artists in two more places in the same movie. And it's subsequently appeared in 1954's *Them!*, the original *Star Wars, The Empire Strikes Back, Return of the Jedi, The Phantom Menace,* plus *Raiders of the Lost Ark, Indiana Jones and the Temple of Doom, Indiana Jones and the Last Crusade, Batman Returns, Reservoir Dogs, Aladdin, Toy Story,* and many more. What surely began as a Foley artist's joke has become sonic litter. Once you hear the Wilhelm scream, you won't be able to ignore it in dozens of films you love.

Sonic trash includes the really generic, bland, or cheesy music you may hear when you're on hold on the phone. When you're forced to wait to get a problem resolved and also have to hear Michael Bolton or Yanni, that's just adding insult to injury.

Sonic trash should not be mistaken for mere noise pollution. A jackhammer is noise, the necessary byproduct of a tool performing a powerful action (not that it's pleasant). But the car with a trunk full of speakers blasting as it rolls by your apartment? That's not a byproduct. It's a choice by the driver to try and broadcast his personality. To most recipients of the message, even if they like the music, it's sonic trash.

You've heard sonic trash on the street in the form of trite ringtones. In 2011, research firm Gartner reported ringtones were a $2.1 billion business. But at about the same time, consumer analytics group IBIS World predicted the business would dry up entirely by 2016. That's because novel ringtones are, for the most part, sonic trash. Music on a phone might have once been considered neat, but most ringtones fail to provide any kind of meaningful, instant information. Your phone might blast the Crazy Frog tune or Lil Wayne's "Lollipop" (one of the all-time most popular) and send a message to the world about your personality and your tastes. But not forever. Pretty quickly, you'll realize that the song isn't helping anyone understand anything better (Lil Wayne's insurmountable swagger notwithstanding). In fact, the wrong ringtones might even create problems in certain situations — say, during a business presentation where you realize you forgot set your phone to vibrate when your hip pocket starts blaring "Baby Got Back."

Some sonic trash can be recycled. Deployed in a different setting for a different purpose, sounds that annoy or disrupt can be harnessed for boom moments.

The company Kids Be Gone markets a device invented in Wales called the Mosquito. It emits a shrill that's been compared to fingernails on a chalkboard or a buzzing mosquito near your ear — but it bothers you only if you're between the ages of thirteen and twenty-five. Humans don't have the ability to hear that frequency until they're teenagers. Then they lose it by their midtwenties. More than a thousand of the devices have been installed in the United States and Canada. Municipalities and private businesses have used the Mosquito to combat everything from loitering to vandalism — the presumption, of course, is that young people are the ones causing the problems. A Mosquito was installed on the grille of a school vehicle in Columbia, South Carolina, to keep kids from gathering in a parking lot after high-school sporting events, and the school district's emergency services manager said it helped defuse fights that would break out there. "Now there's no confrontation at all," Rick McGee told the Associated Press in 2008. "They just get aggravated and leave within a few minutes."

AVOID COMMON SONIC-BRANDING MISTAKES

Heed these principles of sonic branding, and you'll be well on your way to making sound and music a strategic part of your brand. But be sure to avoid the common mistakes that often mire down brand managers who attempt to deploy sound without really understanding how it works.

Don't wait until the end of your creative process to begin considering sound. Include it right from the beginning, when it can inspire everyone and help build consensus on the right communication, emotional payoff, brand personality, or tone.

Sound and music are not only *tactics;* they're part of a *strategy.* Don't settle for unrelated music that fits each individual project. Instead, call back to the essence of your brand with every sound choice. Spending a ton on advertising is too costly a way to compensate for this mistake.

Don't base your music choices on your gut instinct alone. When you just add music that you like and that you think makes the ad better — with no strategy — you'll find yourself endlessly arguing with everyone in your company who has his or her own opinion about the music. There won't be a clear goal or agreement on the values the music is trying to convey. Lots of big brands make this mistake. They'd never develop a visual identity or a campaign or business plan without a strategy; why would they create a sonic identity without one?

Don't make the even bigger mistake of picking music because you think it makes you or your brand seem cool or relevant. If you choose a song because you like some of the lyrics and you don't pay attention to what the song is really about, your potential customers will spot the disconnect in a hot second. They might even tune out your brand forever. Likewise, don't attach an artist to your brand solely because you like the performer's audience. The message that artist puts out to those fans might not fit your brand's story. (Hey, Pepsi, how did that Ludacris thing work out for you?)

If you have something that works, leverage it. McDonald's, Intel, Expedia: outside of the last three seconds of their advertising, these brands are essentially silent. Apple has device sounds and has essentially created its own genre of music in advertising, but its stores, sponsorships, and apps are all silent. These are huge missed opportunities. What if Apple developed the ideas and brought this sonic equity into its public spaces, its products, its sponsorships and employee-training programs? What if it used the elements in a fun way in social media? What if the company inspired artists or the general public to add to the conversation? It's losing out on immense potential brand value.

Don't waste time and money on thoughtless sonic research. The best way to bias your results and make them meaningless is to ask focus groups what they think of the music or the sound. Whether you're scoring films, ads, or products, the moment you take the sound that is normally *unconscious* and ask people to give you a *conscious* impression of what they heard, you can throw the results out. People will start talking about what they like or don't like, which is pointless. You should be focusing on the experience instead.

Avoiding these pitfalls and heeding the principles of effective sonic branding open the way to all kinds of sonic possibilities. When you learn about all you can achieve and exactly where you can go wrong, you can start to develop your unique sound — for your brand, your message, or yourself. In the hands of people and companies that have mastered these principles, sound becomes a fascinating tool for transformation. In the following chapters, I'll give you a behind-the-scenes view of what those sonic transformations look like.

5

Rethinking Possibilities

W HEN I FIRST MET Esther Lee, AT&T's senior vice president of brand marketing, advertising, and sponsorship, she was responsible for telling the story of a company that employed 270,000 employees. But at the time, AT&T itself was embroiled in a massive brand challenge. As the first provider of data and phone service for Apple, AT&T had helped make the iPhone feel magical. But too few people saw it that way. Instead, as is true for all network carriers and providers of all kinds of data and communications solutions, customers tended to notice AT&T when its network failed. In other words, when your iPhone worked, you loved your iPhone. When it dropped a call or wouldn't connect, you hated AT&T. Very few people in those situations remembered that AT&T was the company that made sure they could summon help when their cars broke down in the middle of nowhere.

And never mind that AT&T was the first to help deliver a dramatically new integrated mobile experience. When the company's mobile network was launched, people expected it to immediately work flawlessly. When it didn't, all of the hard work would backfire — people felt cheated.

"We were living at the tip of the innovation spear," Esther says.

The root of the technical problem behind the more widely pub-licized brand challenge could be traced backed to January 9, 2007. That's when Steve Jobs took the stage at the Macworld event at the Moscone Center in San Francisco to publicly talk about the iPhone for the first time. The company and its service partners reportedly spent $150 million on developing the device. AT&T, for its part, had to modify its most advanced cellular platform to become the exclu-sive provider of service for the iPhone, but the device would forever change the way people think about mobile communication.

Smartphone use caused an unimaginable 5,000 percent increase in network demands within three years, from 2007 to 2010. Between 2008 and 2009, 3G data traffic from the iPhone caused a 2,000 per-cent spike in data in and around San Francisco alone. Reports of dropped calls soared. So AT&T pledged to spend two billion dollars to beef up its infrastructure and erect new towers, string more cable, and upgrade servers and all sorts of hardware. Within two years, the company had reduced its dropped-call rate to 1.32 percent, just two-tenths of 1 percent behind the national leader. Repairing the brand reputation would involve way more than copper wire and new cell phone towers.

By October 2009, Esther hatched a plan. The branding agency she had hired, Interbrand, recruited me to join the team. Together we all had to help the people of AT&T get credit for everything they got right, even when customers expected nothing less. One secret weapon would be sound.

What role can sound play in changing the way 270,000 employees feel about the important work they do?

Plenty. "If we can show up where we're normally invisible, there's value accrued to us, and people understand that we're working for them," Esther says.

AT&T, as the first provider of mobile service for the iPhone, was a big part of why the device felt magical. But the company's bad rep for dropped calls had dominated AT&T's story. To change it, Esther and her team needed to radically change the perception of AT&T inside the company's own walls. And she had the notion that sound could be a spark.

During her tenure at Coke, Esther had championed a sonic logo for the soft-drink giant. She came to me looking for the same sort of thing for AT&T — a sign-off, the auditory version of a corporate logo. She knew a sonic logo would make AT&T's brand stickier and more visible as the company worked through a massive rebranding. Studies have shown that a well-crafted sonic logo applied consistently increases the likelihood that television viewers will link an advertiser's message to its brand and forge an emotional connection. Duke University psychologist Wanda T. Wallace, for example, found that "music provides a very powerful retrieval cue. It is more than just an additional piece of information. It is an integrated cue that provides information about the nature of the text in the ad." Besides Coke, Esther's other primary points of reference included some of the ubiquitous sounds we all know — McDonald's "I'm Lovin' It" jingle and NBC's three-note chime, to name a couple. But she didn't want to just put a coat of paint on the outside of AT&T's barn. She and her team decided to incite a broader revolution. And that would take more than a sonic logo.

This is exactly the kind of problem that sound, and an anthem in particular, can crack. It takes rigor. It takes creativity. But the return on the right investment is unmatched. Sound can't make you believe AT&T will never drop a call, because that's not true. But it can tell stories in otherwise silent moments and help a brand get the credit it deserves for getting things right. It can also translate brand values (really the *personality* of a brand) into the relatable language of the sound you hear every time AT&T makes a connection possible. It can play a part in making you feel a multinational, multibillion-dollar company's humanity.

Don't believe it's possible? Think about Apple's Zen start-up sound. Even if you're rebooting because your Mac froze and flushed your data down the toilet, in four seconds that sound quickly takes you from fuming over a crash to beginning a fresh new experience.

The power of Reekes's start-up sound to create these kinds of feelings is the backbone of a narrative in *The Art of Digital Music* by David Battino and Kelli Richards. It's a story from Robert Henke, the developer of the Ableton Live music software and a pioneer of intelligent dance music (IDM). Henke was playing live in Miami for a huge crowd. Just as his set was starting to peak, his Mac froze. The music kept playing, but Henke had no control over his screen. He had to restart. But he couldn't just go silent while he did — this kind of electronic-music event is like an emotional journey for engaged dance crowds. They'd never forgive him if he yanked the sonic rug out from under them. What's more, he had to figure out a way to get back into his music once he restarted without throwing a gigantic kink in the rhythm. So Henke got creative. He fed a simple *thump-thump-thump* beat into his effects processor and looped it while he restarted his machine. The sparse beat created a massive amount of tension — the crowd was begging for a break. Henke had a surprise. He'd fed the Mac's sound through another effects processor, a reverb unit. And when it came back, the simple *thump-thump-thump* exploded into the refreshing, palate-cleansing, blue-sky *waaaaagh!* Reekes's Mac start-up sound transformed a potential dance-club disaster into a mind-blowing break at an unforgettable rave — and club-goers came up to Henke after the show to tell him as much. "I love how failure was integral to my design," Reekes says. "I knew I was composing a soundtrack for failures."

Reekes, like any good sound designer, had started with a problem at Apple and solved it by telling a new sonic story. To tackle a challenge like AT&T's, you need a strong sonic strategy — a rigorously researched and considered game plan for how to use music and sound to bring a brand story to life. It helps you to *feel* the brand. Think of it like a great film score: It supports the story, action, and dialogue. Grammy- and Oscar-winning composer John Williams employs his own kind of sonic strategy to instantly tell you Darth Vader is evil as he emerges from the smoke in his first appearance in the 1977 film

Star Wars. Part of the sonic strategy of Grammy- and Oscar-winning Hans Zimmer for the Dark Knight trilogy involved creating the bat-flap sound to immediately tell viewers that everything they are seeing is part of the story of Batman, even though they don't see much of the actual character for four or five reels. "That sound is enough to let you know you're in the right movie," Zimmer says. "You didn't by accident go into the wrong theater." That's the bar that smart sonic-branding experts and sound designers set: powerful sounds or music that instantly tell a story, trigger memories and emotions and make other senses fall in line. In our case, we wanted people to hear — then feel — AT&T's humanity.

So the team that Esther had assembled inside and outside of AT&T started with an anthem. We were trying to bring AT&T's mission to life — and an anthem, remember, is one of the principles of effective sonic branding. Without it, there's usually no story. From this anthem, we'd be able to craft a memorable sonic logo.

If you're one of millions who've heard the four-note jingle at the end of AT&T ads, you might think we were trying to borrow from the playbooks of McDonald's jingle or NBC's chimes. We weren't. The fast-food chain's "Bah-dah, dah, dah, dah . . . I'm lovin' it" jingle is ubiquitous, sure. But that's mostly because McDonald's has spent billions of dollars on ads featuring the ditty. (You might even remember Justin Timberlake singing it.) If you put a billion impressions behind anything and hire top-tier pop stars to sing it, people are probably going to remember it. NBC's three-chime signature is one of the most recognizable in the world too. But it's aired since 1929, when sonic marks were intended only to notify you of a station's ID, more a requirement than an opportunity. Over the course of eighty-five years, the network has spent billions on media exposure to emotionalize the sound and bring it to life — in songs for golf, football, news, and more. If you've got the budget to buy billions of impressions or if you have decades to instill a legacy behind a sound, by all means, go the NBC and McDonald's route. Esther wanted the effect of those now-ubiquitous sounds. So the team set out to achieve it with a holistic, efficient sonic strategy.

As effectively as sound can tell memorable stories, there's a rigorous, intentional, proven process behind an effective sonic strategy. AT&T's four notes are about so much more than those four notes. There's a larger piece of music behind the sonic logo, an anthem behind the jingle. That's the formula to help any brand get recognition in a fraction of the time and for a fraction of the cost of any one-off jingle or standalone sonic signature. It took just a little more than a year for our sonic campaign to really sink in. If you know AT&T's four notes, you've probably heard its anthem in one form or another too, whether you realize it or not. AT&T's sonic logo is the culmination, not the beginning, of how the team and the very sharp folks at AT&T used sound in a campaign to tell a new chapter in the company's story. By the end of this chapter, you'll start to see how any company, big or small, can use sound to tell its story.

Our work for any company starts with an investigation. Like scientists tracking a rare species, we embark on a months-long mission to listen to all of a brand's sounds in the wild: in stores and in TV and radio advertising. Then we turn into forensic investigators, poring over employee-training and investor materials, web pages, everywhere any customer or stakeholder comes into contact with the brand. Then we run our client's competitors and partners through the same audits. We even include a handful of brands from outside our client's immediate circle. We often study hundreds of samples from dozens of companies, then we chart all the results based on a variety of criteria, including what sounds each company owns and which sounds are up for grabs. We are looking for the territory where no sound has gone, searching for the creative opportunity to stand out.

When it came to AT&T, the company lived in a remarkably homogenous and uninteresting sonic landscape. Technology brands were often too predictable in their use of computer-like sounds. And yet, most of the songs licensed by each brand didn't sync with the brand's personality. They used pop tracks that could have worked just as well in spots for chewing gum or denture cream. Nothing stuck out and told a story. There was very little differentiation. And with the exception of a couple of companies, almost no one consistently used any type of sonic identity or clear strategy. Verizon's sonic elements in its advertising didn't sound much different than anyone else's. T-Mobile had a strong sonic logo but no anthem, and it used the logo only in the last three seconds of its ads.

Most companies took a slapdash approach to using their sounds, music, and voices. The sonic elements they did own merely complemented pictures or seemed to be chosen because they sounded cool. To understand how astounding this lack of strategy is, consider how much top global brands invest in all sorts of other brand research. Companies in almost any other business category use a visual-identity system that includes visual logos, colors, shapes, and rules on how to apply them to represent a brand's values. You'd never expect, say, a kids' cereal or a pack of toilet paper to come in a black box. When Yahoo CEO Marissa Mayer was at Google and in charge of the look, feel, and function of the search giant's homepage, she tested forty-one shades of blue with users and analyzed click data before deciding on a specific permanent hue. And yet, Google has no distinguishing sound. In 2008, once-revered designer Peter Arnell convinced Pepsi to spend a million dollars on redesigns of its logo. In internal leaked documents, he referred to his design by the name Breathtaking, suggesting it was nothing short of a force of nature. He unveiled a new packaging design for Tropicana Pure Premium orange juice that year too, which went over so horribly with consumers that it was eventually replaced with the old version — the whole thing reportedly cost Tropicana parent company PepsiCo millions. So how come companies who spend millions on visual logos and design don't apply even the basic amount of rigor and investment to sound? Our goal was to help AT&T build the triumvirate of storytelling: visual, verbal, and sonic. We knew that there was a huge opportunity for us to introduce AT&T to a deeper kind of power of sound. When a brand strategy and a sonic strategy pair perfectly to tell a true story, the result isn't just good advertising; it's something like a cause or a religion. Plenty of people have worshipped it in the form of Apple.

As part of our process, we decided the best way to translate AT&T's humanity was to use sounds that felt handmade and imperfect. We'd take the focus off technical perfection. Sound and music would help reassure customers that there were relatable humans behind all of AT&T's recent innovations. The team dubbed our sonic strategy the Sound of Humanity, an extension of AT&T's Rethink Possible theme, and it became the basis for every sonic experience you still have with AT&T. It's the guide to what devices should sound like, what music plays in retail stores, what bands AT&T considers sponsoring, what messages accompany advertising. It's what the brand sounds like everywhere.

As with many large companies, AT&T provides a huge variety of products and services and has an enormous ecosystem of customers and employees and a lot of different needs that have to be met. We needed an anthem to tell AT&T's sweeping story, one that resonated with cell phone customers, enterprise clients, the salespeople in the field, and all the scientists in the lab. AT&T's target customers are, essentially, two out of three Americans. And they all had to hear themselves in AT&T's sounds.

The most basic vocabulary in this process for any brand involves instruments and sounds and the feelings and emotions they trigger. There's no codified law on this stuff—more like case law, at best. This is the part of what I do that goes beyond science. Our sonic lexicon is more like a combination of acquired wisdom from thousands of hours of playing and recording virtually every instrument. For the sake of this story, see the following table, which will help you understand the basic kinds of instruments and sounds and their corresponding emotions.

Instrument	Emotions
Strings	Warmth, scale, or scope; passionate, uplifting
Horns	Power, elegance, impact, importance, strength, honor, bravery, and heroism
Synthesizer	Modern, forward-thinking, evolutionary
Piano/percussion	Heartfelt, emotive, personal, driving energy, velocity, or anxiety
Drums	Driving, motivating, primal, communal
Electric guitar	Power, youthful energy, rebellion

Before we could translate anything into the language of sound, we all had to speak AT&T. The company values inventiveness and is always endeavoring to come up with unique ways to connect people who want to communicate. Anyone who's ever heard the soundtrack to *American Beauty* or *A Beautiful Mind* or the syncopated piano riff from Coldplay's "Clocks" has the sound of *inventive*. It's plucky, fresh.

AT&T is also purposeful. Its pursuit of innovation isn't whimsical. It's relentless. And people look to AT&T's technology to get things done. From a musical standpoint, *purposeful* has a certain propulsion to it. It sounds like the Penguin Café Orchestra or Philip Glass.

Curious describes the quirk in AT&T's personality. *Curious* sounds like Danny Elfman's score to *Edward Scissorhands* or the opening to *The Simpsons.* It sounds a little like *inventive,* but it approaches the line of adventurous or weird.

The toughest of all of the company's attributes was *open.* For decades, technology like AT&T's was proprietary and closely guarded. But tech circles had become more oriented toward collaboration (*open* as in *open-source*). Musically, *open* sounds like *simplicity* and leaves lots of sonic white space. It's a string quartet or an acoustic guitar recorded in a cavernous room. It's that group-sing dynamic. Everyone's invited.

That last one is precisely where we went wrong with the first version of the AT&T anthem. Cut to my Midtown Manhattan studio. Esther Lee and I are leaning on opposite corners of the studio console with key AT&T executives and creative directors from their ad agency BBDO in the room waiting for us to wow them with new music for the brand. I reach for the white button on the console and push Play. Up swell the guitars, organic-sounding synths, and chorus of voices — the beginning of AT&T's new anthem. We'd sunk about five hundred hours into exploring dozens of different thematic ideas, refining them, developing the right mix of sounds and melody. We'd employed twenty-five musicians who played strings, accordion, brass, piano, bass, drums, guitar — I played some keyboard. We had fifteen writers involved, including myself, and I produced it. We'd laid down a pretty big swath of stuff for multiple tracks and sketches of these ideas. The theme we had decided on was an optimistic pop anthem with group-sing vocals, seasoned with just a bit of Brooklyn band crunchiness. Think Arcade Fire or the Lumineers (whose song "Ho Hey" Microsoft would later use in a campaign for its social search engine Bing). We loved it. There was a nod to every brand attribute in there, plus all the right colors and moods. Or so we thought.

During the minute and a half of play, I'm scanning the room for smiles and toe taps, looking for the nods I'm sure I'll get. But I'm not seeing any of that. Instead, I catch a few of the guys from the larger team shooting one another uneasy looks across the room. They know what's coming.

When the music ends, there is silence. Finally, the executive in charge of consumer advertising speaks up.

It sounds too happy, he says.

I suggest how *happy* can also be perceived as *open,* one of AT&T's core attributes, the one that's newest to them.

Another department head adds that it feels too much like smiles and rainbows, and he isn't convinced that most people will buy that sentiment.

Murmurs fill the room: *Hmm, yes. It does seem a bit sunny . . .*

My reflex is to keep fighting. But it won't matter; we don't have a buy-in. This is supposed to be a victory lap. But just about every executive crammed into our cold, dark studio thinks everything he or she is hearing sounds too optimistic and not inventive or purposeful enough for AT&T. "It just doesn't speak to the important work we're doing," someone says.

My jaw dropped. What the hell happened?

Translating a brand into sound is tricky business. It's like capturing a person in all of his or her fullness and complexity. "It's a very subjective area," Esther says. In a company as large as AT&T, lots of stakeholders have to feel good about it. I've learned to pay attention to everyone's emotional reaction. They all know what their brand should feel like. They're all qualified because they have ears.

It takes only one poor translation for a brand's sound to feel like bad cosmetic surgery — fake, desperate, and a little pathetic. We'd misinterpreted *open* just enough to create a mismatch between the emotion the music triggered and the message AT&T wanted to send. Its mission was all about connecting people. AT&T's technology platform is the lifeblood of many technologies that people couldn't imagine living without. It was the first network to power Apple's iPhones and iPads. And whether or not you've suffered dropped calls, the network still connects millions of emergency calls — or even just really important calls — each year, on landlines and cellular networks alike. Sound could help AT&T get credit for experiences like those. What it couldn't do, as I learned that harsh day in my studios with AT&T execs, is tell a lie.

The moment the executives filed out of the cramped control room and headed off to the airport to return to work in their own corners of AT&T's universe, we got back to work. We knew we had to throw out everything and start again, even if we weren't immediately sure what we'd missed.

After we tore down the whole anthem and started sifting through the pieces, we began to see which principle of sonic branding we'd trampled on. As abstract as all of this business might sound, it becomes incredibly concrete when you get something wrong. The problem was that our music didn't tell the story of the brand's purposeful work, its relentless innovation. AT&T had taken lumps lately. If the music smiled through the pummeling, it wouldn't ring true. Mostly, we'd taken for granted that sound and music could not make a lie believable. We were using sound to tell the wrong story. Our sound wasn't telling the truth.

Fortunately, AT&T had a true story worth telling. We just had to translate it properly. David Lubars, the chairman and chief creative officer at BBDO, AT&T's primary ad agency, gave us a fantastic tip: he told us we needed to find the musical version of the Rosetta stone. He suggested to us that if we were going to deliver on the *humanity* of the brand, we might want to become musical deconstructionists. We should create an anthem that could be reduced to anyone's personal melody — a few stripped-down tones — that we believed represented the soul of the brand. Everything would be built on that foundation. We threw out the rule book when it came to all of those traditional instruments, like horns and strings. This time, to capture the more human-sounding, curious, purposeful, inventive, and open attributes, we rented some old, broken, less glossy instruments. We went with a beat-up glockenspiel, an old upright piano, a 1970s-era clavinet played by Stevie Wonder the week before, and bagpipes (to the best of my crew's knowledge, that's the only time anyone's used bagpipes in a corporate sonic logo, but it worked). Anyone who plays the bagpipes will tell you it's one ornery instrument. You have to coax it to do what you want, and it's almost impossible to get it right on the first try, let alone to make it fit into an odd assortment of instruments. That's the kind of imperfection that ultimately helped capture a new layer of AT&T's humanity and the idea of relentless innovation that the company practices.

Fast-forward through a couple of weeks of furious studio work.

David and I met at BBDO and together we repitched the music to VP of brand identity and design Gregg Heard (yes, the guy in charge of sound at AT&T is named Heard). Gregg bought into it and brought it to Esther. She was thrilled. But she still wanted her sonic logo. "It's already there," I told her. "I'll prove it to you." Then my team and I stripped away the layers of the anthem until all Esther could hear were the four notes. Suddenly, it was clear that every piece of the anthem included some form of those notes, the ones you hear now at the end of AT&T ads. Try it yourself. Listen for them not only at the ends of ads for AT&T but in the phones it sells, the stores it owns. Everywhere. That's the true test of an anthem. If it can't be distilled into a simple theme that will echo out across the whole sonic story in all kinds of settings, it fails.

The most effective sonic strategy starts with a big story. Then you distill it until you get to the moral of the story. The moral can adapt with the changing times and fit all kinds of moods. It can evolve just as the famous visual logo or even bottles for Coca-Cola have evolved, moving in sync with popular culture and anticipating the needs of future customers.

After we played the signature for Esther and then demonstrated how it worked in different contexts and for different audiences across the brand (consumer and business-to-business ads, devices, sponsorships, and spaces, among others), she brought samples of the anthem and sonic logo to the wider team, including the executive leadership. They reviewed the work in the context of advertising, products, and services. We showed a mockup where the anthem played as an executive took the stage at a global telecom conference. That got a few hoots and hollers from the larger team members, who appreciated the potential rock-star moment. This time, we knew we'd nailed it.

Before the final version shipped, though, we still had to work through a few adjustments. Specifically, there was one iteration of the anthem that became the default ringtone for the Nokia Lumia line of phones, which ran on AT&T's network. The team wasn't ready to sign off on our first choice, and they shot it down. It still wasn't clear and simple enough for them. Why would some of the top brass at a Fortune 500 company take the time to critique a ringtone? Nokia phone ringtones alone are heard 1.4 billion times a day. It's the iteration of the sound that makes people most aware of the brand and what it gets right. "We think of the sonic logo in terms of the graphic logo," says Gregg Heard, who presented the final product to the executive team. "It has the same position within the company." If you use that kind of logic, why wouldn't the company's decision-makers want to sign off on the sonic logo? "We thought it was at the same level of importance," Gregg says.

Based on the executive input, we went back and created a bunch of default ringtones and recommended a couple of new choices, which almost everyone liked.

Those ringtone versions of the AT&T anthem, in several varieties of moods and styles, now come loaded on hundreds of thousands of phones. But this is just one way that the sonic brand is being used these days. If you've watched any major sports or pop-culture event on television lately — from the Super Bowl to prime-time sitcoms and reality shows — you've probably seen an AT&T ad depicting one of the myriad ways the network makes things possible. And at the end of every one of these optimistic scenes are four plucky, determined, sparse notes followed by the sound of fingers snapping. It's the sonic logo Esther and her team set out to find and that we distilled from a sweeping brand anthem. You might also hear those notes playing at AT&T-sponsored venues at music festivals like SXSW or at the stadiums the company sponsors. It's woven deep into the transformative, surprising experience at AT&T's flagship innovation store in Chicago. In all of these ways, and in new versions of the anthem and logo that we are still working on with AT&T, the sound of the company finally helps gives credit where it's due: to the 270,000 employees who help power all of these human connections that weren't possible just a few years ago.

You don't have to be AT&T or use Stevie Wonder's former clavinet or even sink five hundred hours into a custom anthem to find your own sonic strategy. There are tremendous opportunities for global businesses to use sound, but it's a powerful tool for small companies and brands too.

If you own a small mom-and-pop business, you can do a number of simple, inexpensive things that make a world of difference. Creating a sonic experience doesn't necessarily take a lot of money. The right sound can often be the most leveraged and cost-effective investment a business can make for a better, more distinct customer experience, whether you own a pizzeria, a pet store, or a flower shop.

What about choosing your in-store overhead music? There of course are services that can do this for you (as long as *you* tell them what to play). But with some careful thought about what your brand is all about and how it's different from the competition, you can better score your store. Are you more about service? Or value? How about convenience? How does that measure up to your competitors? What music do they have in *their* stores? If they are playing nothing or, better yet (for you), the same top 40 radio as everyone else, you've already won. They're going to be easy to beat. Most overhead music just takes up space and tries to entertain, but you know that it's really the way to speak *your* brand with your own voice. You can achieve this by playing deeper cuts by the same popular bands or nostalgic hits targeted to your typical demographic of customers that you probably know well.

If you own the flower store that is all about personal service, what about a modern coffeehouse mix to slow people down? You could step it up to a medium tempo on Valentine's Day, when you want people to move through the store a little quicker. Not musically inclined? No problem. I bet you know a customer who is. Get him or her to help you pick a playlist and post a notice letting everyone know about the guest DJ. Once you've made these choices, you are no more than an iPod and a couple of speakers away from helping people *feel* your brand, and you'll start instantly beginning to build an emotional connection with them. Will the food taste better, the clothes seem more interesting, or the flowers smell sweeter? You bet.

Here's another idea to consider: What's the sound of your store's front door? Is it welcoming? Is it a big old squeaky door, or does it *whoosh* open? The sound of that whoosh welcomes people and sets a mood in an instant. Is there industrial carpet on the floor, or wood? Is the sound in the space echoey or warm and inviting? Be sure that all these choices add up to a singular, cohesive experience — they should make your business a place customers recognize and love to visit. If you sell antiques, you might use an antique cash register to ring up sales. What if your candy shop piped in the sound of candy being made, turning a neighborhood store into something more like Willy Wonka's chocolate factory? Simply put, how could curated sound make the experience of your small- or medium-size business stand out?

By itself, sound might not save mom-and-pop stores throughout the country in an age of superstores and e-commerce. But it sure can make employees happier, more productive, and more relaxed. And helping customers connect to your business could be the beginning of a revolution.

AT&T, though nowhere near what anyone would consider *small*, is a business built on customer connections. And it accomplished as much with sound in twenty months as other comparable brands have achieved in decades. Six months after we launched our first sonic campaign with AT&T, the sound was almost as recognizable to consumer study groups as the AT&T globe logo that'd been around since 1983. After fourteen months, a quantitative study found that most people recognized the AT&T sonic logo; almost as many as the number of people who recognized the NBC chimes, which were first heard in 1929. More important, the number of people who recognized the sonic logo as belonging to AT&T without any prompting was twelve times greater than the number who recognized the sonic identity for Coca-Cola, which launched in 2007. (Can you even hum it?)

Whether it's AT&T versus T-Mobile or a cornerstone bodega versus a big-box supermarket chain, sound can accomplish in a few notes what millions of dollars in visual advertising and radio and TV ads can't. The perfect music for a brand can make an emotional connection to attract customers and keep them coming back.

6

Amplifying Messages

I N THE PAST SEVERAL YEARS, a change has swept across the American middle class. It's led by second-generation Hispanic and Latino Americans, who numbered almost fifty-two million in 2011 — they're the largest and youngest minority group in the United States. And second-generation Hispanic Americans, people whose parents were born elsewhere but who moved to the United States, have higher incomes than the previous generation. More graduate from college. More own homes. Most speak fluent English and Spanish, according to Pew Research. The research also makes it clear that this is the demographic that will have the biggest impact on the character of America in the near future. In a 2011 Pew study, 47 percent considered themselves "typical Americans." "We are the middle class," says Ruth Gaviria, the executive vice president of corporate marketing for Univision Communications International, home to Univision and its sister television and radio networks. Univision commanded the largest audience of Spanish-language TV viewers worldwide in 2013, according to Nielsen.

As Gaviria is well aware, from a marketing standpoint, there are challenges in speaking to this new group of people. Within this group, people are incredibly diverse. People with Mexican heritage make up the largest amount of the Hispanic and Latino population,

33.5 million in 2011, according to the U.S. Census, or 64.6 percent of the total Hispanic population in the United States. But there's no agreement on how they refer to themselves culturally. Most identify with their parents' country of origin, with lower percentages identifying themselves as Latino or Hispanic. Hispanics are also assimilating rapidly. About 26 percent of second-generation Hispanics have a spouse of a different race or ethnicity, according to Pew. About 60 percent of Univision's audience identify themselves as being of Mexican descent, but there's a healthy mix of Colombian, Dominican, Puerto Rican, Spanish, and more. "We're a whole bunch of awesome *others*," says Ruth, who herself is from Colombia.

The diverse group is, however, united by a set of distinguishing values. "And our cultural values are now being threaded through America," Ruth says. Values such as hard work: 78 percent of Hispanics and Latinos say they believe most people can get ahead if they're willing to work hard. (By comparison, only 58 percent of the U.S. population believe this.) This isn't cheap, generic, unskilled labor shouldered by an invisible underclass — it's a value that's more about the nobility of any work with a purpose, a work *ethic*. Family is another big value too, especially for the segment of the population that Univision speaks to, Ruth says. By her and other insiders' accounts, the network was stuck in the values of the 1950s. And for a long time, that was enough to keep it in line with what was important to its audience, especially in tight times. "When the economy crashes, that's exactly where we need to go," Ruth says. "We have no money? Family's important."

It's nearly impossible for non-Hispanics to understand the place that Univision holds in the hearts and minds of American Hispanics. The media landscape for the traditional English-speaking broadcast networks has radically changed. In decades past, huge audiences used to watch prime-time shows every night and then share water-cooler or coffee-klatch conversations about their favorite programs the next day. Today's non-Hispanic audience is splintered among hundreds of channel options and online, on-demand entertainment and news sources vying for their attention.

Univision, by contrast, is still synonymous with a unified sense of family and trust. Most of Univision's audience grew up with the network perpetually on the television in their homes; Univision anchor Jorge Ramos remains the Walter Cronkite of Hispanic news media. As Ramos goes, so goes the Hispanic viewpoint of the nation.

For years, the network had maintained an enormous lead over the next highest rated Spanish language channel, Telemundo. Univision had mastered the art of giving viewers the programming they said they wanted. But something was missing. Univision Communications CEO Randy Falco, along with Ruth and others, had a sense the brand could deliver much more. "We make entertainment easy . . . ," Ruth says. "But that's not what this network really stands for."

By 2010, the network started to feel the foundation of its viewership shift under its feet. By 2011, it was transforming into the mainstream in a quantifiable way. One week in September, Univision beat the ratings among eighteen- to forty-nine-year-old viewers for ABC, NBC, CBS, and Fox, a first for any Spanish-language network. (That week, Telemundo averaged 810,000 viewers to Univision's 3.8 million.) Ruth, along with Falco and others, started to realize that the network had a chance to align itself with this changing story. Or, alternatively, it could get left behind.

We saw with AT&T how sound can have a profound impact on a business in transition. AT&T was instrumental in revolutionizing the way we use our mobile devices, but when Man Made Music linked up with the telecom company, AT&T was in danger of losing its place in the story. It was becoming known for dropping calls instead of connecting them, even though it connected calls way more often, and our sonic challenge was to help them turn that perception around. For Univision, the cultural stakes and the opportunities were even greater. Their challenge to us was to help them amplify the positive brand associations they'd already established and to unify their audience as part of a greater movement. Univision has provided a voice of unity and served as a hub for a shared set of values for decades. Now the Univision corporation faced the enormous challenge of aligning itself with the right notion of community for a generation that was itself still grappling with its own identity. Its audience still felt a sense of responsibility to cultural heritage. But they had evolved beyond walled-off barrios and enclaves and were deep into a kind of reverse colonization. These are people who'd grown something from virtually nothing and developed a healthy adaptability along the way — an American value if there ever was one. "It's a movement," Ruth says. "But I needed to let the world know that it existed."

Keyed up in just the right way and played at the right moments, music in the form of a distinctive anthem — one that could be distilled into all sorts of logos and brand-navigation sounds — could help Univision instantly, frequently, even subconsciously relay a signal back to its viewers whenever they came in contact with the company, a signal saying it was alongside them and even occasionally leading them on their journey toward the cultural mainstream. At the same time, it could take credit for all of that loyalty its audience members already felt. Univision could use sound to cue up nostalgia even while pushing into the future, the same way Ford did with the sound of the exhaust on its 2008 Bullitt Mustang. With the help of the right sound at the right moments, Univision could align with the change sweeping over its audience and the country as a whole.

Whenever the network held focus groups to measure brand awareness and regard — marketing-speak for "trust" — Univision always "broke the scale," as Ruth puts it. It was a sign of the potential the company had yet to realize. There's a story Ruth tells that speaks directly to the level of trust Univision's audience put in the company. At about the same time Univision started to see its ratings challenge the big networks, Ruth's assistant got a call that she knew to patch right through to Ruth. On the line was a young mother of Mexican descent who didn't speak fluent English and was trying to figure out how to get her elementary-school-age child enrolled in public school in Texas. She'd found a number for Ruth's office through Univision's website. Desperate, she'd dialed it.

"This was a woman full of anxiety. She had never registered a kid. She didn't know how the process worked," Gaviria says. "She needed help. I stopped everything I was doing." She's done this lots of times, actually. She's fluent in Spanish, which can quickly tamp down stress for someone who doesn't speak English well. "So I'm taking all of the data — where they live, how many kids do they have — and I'm on the Internet and I'm looking at all of the elementary schools," Ruth says. She found the right one and gave the woman a number to call that had a Spanish-language option. Then she gave the woman her direct office line. And finally, she gave her one more number: the woman's local Univision station, which she could call if she had any more questions.

Strange as contacting a TV station for life advice might sound, this wasn't such an unusual call for anyone, even an executive, at Univision to receive. "Taking a phone call like that is sacred for us. We have all lived those moments," Ruth says. But in the context of its growing ratings and the demographic shift happening beneath the network's feet, the call took on new meaning.

Executives, including Ruth, realized they were sitting on a vast well of trapped value. "One thing that we weren't capitalizing on was the brand equity," Ruth says. "We had permission for more."

Randy Falco's idea, brought to life by Ruth and others, involved shifting the focus from the content to the viewers themselves. There would always be bread-and-butter shows and specials on Univision channels — telenovelas, soccer, and the like. That's what audiences always *said* they wanted to watch. But the broader strategy involved programming that spoke to viewers' emerging identity — their emerging, distinctly American demographic. "That bicultural audience is happy to watch the telenovelas as comfort food, but they also like to watch the Kardashians," says Linda Ong, whose branding firm

TruthCo helped Univision (and me) find its voice. By the time I got involved, in fact, everyone was clamoring to hear what that voice should sound like.

Music for Univision was far more important than it was for a typical network. "Music has always been very important to the culture, period. It's one of the major cultural tenets that we keep," Ruth says. But it could have a far greater impact. It could help cement a set of shared ideas and galvanize the groups of people who hold those ideas to form a kind of virtual nation — even if the people themselves didn't share a common national heritage. "So we started creating an anthem to validate a nation," Ruth says.

Linda and Ruth had decided that Univision could help unite a community with a common set of values and the numbers to influence America's future. They needed a sonic strategy and the musical language to deliver on that story.

Typically when we get assignments at Man Made Music, we survey the sonic landscape of a brand's competitors. But not only did Univision far outpace its closest competitor, it wasn't even trying to compete with other Spanish-language networks. It was aiming at the big four: NBC, ABC, CBS, and Fox. But really, it was trying to align itself with a cultural movement. So instead of surveying other television networks, we had an excuse to study the musical landscape of Hispanic and Latino culture as well as the role of music in great historic movements.

Almost every major cultural shift worth noting in history has been transformed by a unifying piece or genre of music. Music is often the tool that expresses a shared set of beliefs, ideas, and morals — the fabric of societies as described in 1893 by French sociologist Émile Durkheim's *Division of Labor in Society* (the idea of the collective conscience). Whether it's based on nations, cultural revolutions, rock bands, or computer operating systems, our connection to groups of like-minded people helps us define our identities in the world and motivates us to action. When the expression of that connection takes the form of an anthem, amazing things happen.

Take, for example, "The Star Spangled Banner," whose lyrics were adapted from Francis Scott Key's poem "Defense of Fort McHenry." It was written in 1814, seventy-nine years before Durkheim published his theory. (To be fair, it didn't officially become the national anthem of the United States until 1931.) But it wasn't until Key's imagery was set to a melody that it found its deepest historic resonance. With music, "the bombs bursting in air" became a literal boom moment, first and foremost uniting colonists who would become citizens of the United States of America.

New nations or those fresh from turmoil immediately turn to music to unify their people and bind together feuding factions. Less than a month after the end of its violent struggle for independence from Serbia in 2008, Kosovo held a national contest for a new anthem. In years prior, officials had played Beethoven's "Ode to Joy" at ceremonies, but it hardly galvanized the country — Kosovar Albanians still sang the Albanian anthem. But, as a government source told the *Hurriyet Daily News,* "If we put the anthem in Albanian, the Serbs would not see that anthem as their own." In June 2008, the newly formed Kosovo decided on "Europe," composed by Mendi Mengjiqi. It sidestepped controversy with Albanians and Kosovo's Serbian minority by doing away with lyrics altogether. Language itself carried too much conflicting meaning for an anthem meant to unite. But the language of music could uniquely convey the emotion that most citizens wanted to share without the divisive subtexts of language. Kosovo's national anthem is a good example of how music can navigate the treacherous and unsettled national waters that words often can't.

A lyric-less anthem became the solution for Spain too. The Spanish Olympic committee planned to have Plácido Domingo sing part of a new anthem at the 2008 games in Madrid, but they scrapped the idea a week before the event when controversy arose over whether sections of the anthem should be sung in Basque, Catalan, Galician, and Aranese — the various languages spoken in parts of the country that still consider themselves virtually independent. In the end, Spain's anthem, "Marcha Real," like Kosovo's, went wordless.

Timothy Garton Ash, a columnist for the *Guardian*, put it best when writing about the anthems of Spain, Kosovo, and others in 2008; he called successful national anthems more than tokens of nations. They're "part of the nervous system of a living political community," he wrote. But music need not have political implications to unite a group of people under a common identity. In the same way an anthem becomes more than the token of a nation, certain bands or artists, like a great orator or charismatic leader, can become central parts of cultural nervous systems. The right unique music at the right time can secure a culture's or movement's place in history while allowing it to reverberate through generations. And the artists' legacies are swept up in all of this.

The Grateful Dead arguably started the freewheeling, rock 'n' roll, post-hippie, jam subculture. Its members shared their lifestyle and spirit not only through music but in the creation of massive experiences where communities could unite and share stories, goods, bootleg tapes of other shows — each one was unique — wares, drugs, and bodily fluids. The members of the Grateful Dead had a vision for their sound design. Their sound worked for that moment in time not just because they were accomplished musicians who could impressively and passionately noodle on their instruments for extended periods. It was also because their audiences were looking for that kind of extended experience too. The Grateful Dead managed to prolong an open, free-spirit aesthetic even as the country as a whole experienced a loss of innocence. They were the safety net for hippies who weren't quite ready for the big jam to fade. And since every show was slightly different than the previous — and someone was always rolling tape to make sure that moment lived on even longer than the marathon concert itself — the earthy, festival-following modern Dead-head devotees were still singing along to *American Beauty* bootlegs long after bandleader Jerry Garcia died. The group's narrative and lifestyle was carried on by modern torchbearers such as the String Cheese Incident and Phish in the 2000s.

On the other end of the spectrum, there's the Sex Pistols, who were widely associated with the foundation of punk rock, the counterbalance to the late-1960s hippie-driven musical movement. Whereas a Dead song could go on for tens of minutes, a punk-rock song usually chopped off at two or three, tops. Many of the musicians couldn't play more than three chords. But they poured on the angst and confrontation and designed a sound that united a crowd of hard-core individuals who felt the same sort of angst, even if they weren't so sure where to direct it. The bands of the punk-rock era volunteered themselves as targets because they so desperately related to the feelings of wanting someone to blame for the political and socioeconomic situation they'd been handed. The Sex Pistols tapped into that complex sentiment with percussive guitar, drums, and bass and a nasty, whiny style of singing that defied melodic rules and sounded more like sonic rioting than anything else. The formula didn't die with the breakup of the band or the death of its short-lived, charismatic prototypical punk-rock bassist Sid Vicious. When barely knowing how to play an instrument or defying the idea of melody became a played-out trope of punk rock, the Ramones, the Clash, Iggy Pop, and the Dead Kennedys actually injected technical chops into the equation but then wrote songs that challenged the parameters of pop. The later pioneers of punk rock figuratively snuck into the mainstream — they'd write catchy hooks in songs such as "London Calling" and followed pop-music structures in tunes such as "Blitzkrieg Bop" but included acerbic social or political commentary as a secret signal to their core fan bases.

Grunge was what happened when punk rock threw open the gates, with Nirvana as the poster band for a lifestyle that, like punk, found its way into fashion, art, and even the mentality of a young generation of "slackers." Despite the name bestowed upon them by those who didn't quite understand their anticorporate mentality, this generation produced a rich array of comedy, films, and literature that defined the 1990s. They eschewed capitalism and corporate anything even as they redefined mainstream and became stars themselves.

A cultural movement or milestone certainly doesn't have to be about music in order to be galvanized by it, though. The civil rights movement was bigger than any song or single anthem, but it is anchored in a set of emotions and experiences first shared in the language of music: freedom songs. Bernice Johnson Reagon, a singer and civil rights activist and former member of the Student Nonviolent Coordinating Committee, said on NPR's *Talk of the Nation*:

> I don't have any sense of the civil rights movement existing without the singing we did in marches, in mass meetings and in jails. There is no separation. For me if I hear a program about the civil rights movement ... if you're not listening to some of these recordings, I feel you've missed an opportunity to hear the energy and voice and understand the articulate voices of the masses of people who stepped out of the old ways of being and got *in* the way to change and give us a new situation.

Reagon's daughter, Toshi Reagon, a singer-songwriter herself, added that the music "was also a way for individuals to actually not be alone, a way to really survive some, you know, horrible situations and some uncomfortable situations. By bringing your voice and bringing your voice inside yourself and with others, it created a power and a way to exist in the world."

Freedom songs were often on-the-fly adaptations of spirituals sung by enslaved Africans in America as expressions of faith, shared experience, and dreams of freedom. The perfect moments of musical expression in spirituals offered hope in a miserable existence, and they reverberated across decades and eventually around the world. Spirituals and freedom songs inspired a whole genre of gospel music. The hope captured in the gospel song "I'll Overcome Someday," written by Charles Tindley in 1901, for example, inspired the internationally recognized protest anthem "We Shall Overcome" in the

1960s. That song's emotional themes inspired everyone from president Lyndon B. Johnson, who uttered the phrase in support of the by-then full-fledged civil rights movement in a March 1965 speech before Congress, to Martin Luther King Jr., who used the phrase as the title and refrain in his speech on March 31, 1968, just four days before he was assassinated.

In recalling how gospels and spirituals helped form the soundtrack of the civil rights movement, King himself wrote in his 1958 book *Stride Toward Freedom: The Montgomery Story,* about the Montgomery bus boycott that began in 1955 after Rosa Parks famously refused to give up her seat: "One could not help but be moved by these traditional songs, which brought to mind the long history of the Negro's suffering."

King's oratory style was songlike — as a preacher, he learned to use speaking patterns that harnessed the emotional power of music. He frequently quoted and made allusions to spiritual music, using familiar song lyrics as emotional triggers in his most memorable speeches. Orators have long noted the particular power of King to summon blissful, uplifting endings to his speeches. The lyrics of a spiritual, "Free at Last," are the crescendo of King's "I Have a Dream" speech, considered his most influential. King might have created history's most powerful boom moment by tapping into the power of a song lyric that instantly conveyed a place in time, a rich history of a specific set of disenfranchised people, and hope for their future — all in three musical words. In his emphatic and melodious tone, he shared his vision in which "Jews and Gentiles, Protestants and Catholics, will be able to join hands and sing in the words of the old Negro spiritual, 'Free at last! Free at last! Thank God Almighty, we are free at last!'"

My team and I entered into our work for Univision with King's grasp of the power of music and song firmly implanted in our minds. We'd even had a chance to touch a piece of it ourselves. In January 2008, David McKillop, a senior executive at the History Channel, was co-executive-producing *King,* a documentary on Martin Luther

King Jr. He came to me with an idea that would honor not only King's historic influence but his relevance to the present and future. Visually, David planned to present new interviews with luminaries from art, politics, and music on the impact of King. But most of the feature would be culled from historical footage. "You've already seen every frame of MLK footage," he told me. "They're not making any more. . . . So the soundtrack will be key. How do we tell this story with music? What should the soundtrack be?"

I fell back on the idea of spirituals as the score for the civil rights movement, and then the team started to consider how we might carry that music through the entire story and into the future to create a new King experience from very familiar imagery. We talked about how the language of spirituals was threaded through King's speeches. In more recent times, we realized, U2 had borrowed the phrase *free at last* for the climax of the band's own song "Pride," which was an ode to King. That's about the time we began discussing with Susan Werbe, another executive producer on *King,* and the NBC production company Peacock Productions the idea of having a contemporary artist cover "Pride." It would be the perfect way to harness the emotion of King and all he stood for and carry his legacy into the future.

We approached John Legend, who signed on immediately. We decided, with Saevar Halldorsson, then creative director at the History Channel, to record a stripped-down version of the U2 classic — just John singing and accompanying himself on the piano. We went to work at Legacy Recording on West Forty-Eighth Street in Manhattan and nailed it in about an hour — just four takes.

After the recording and the documentary aired, things got really interesting, and the hunch that this musical idea could amplify a historic message and send it into the future proved true. Colleges played the song at graduations. John put it on the UK release of his album *Evolver.* Producers interspersed the song with bits of speeches by Barack Obama for the 2008 album *Yes We Can: Voices from a Grassroots Movement,* a fundraiser for Obama's campaign. And it appeared on the 9/11 memorial tribute album *Ten Years On*. "It took on a life of its own," John says. "That's the evergreen nature of a great song. They can be resurrected. They can be covered. They can find new relevance due to changing circumstances in history."

When you find the right material and use it strategically to craft a boom moment, it can echo across ages and turn up in places you never expected.

Our understanding of how national anthems unite seemingly disparate groups of people — not just as nations but also as cultures, subcultures, and historical movements — set the stage for our work with Univision. It was a brand, but Univision could align itself with a cultural movement and unite people from several very distinct places in the world under one common set of values. And music could be the glue. My colleagues and I set out to create an anthem for Univision that would anchor the network, clearly align it with the evolution of its diverse Hispanic and Latino audience, and scale across media the company hadn't even launched yet.

Pretty quickly, we realized that we couldn't unite this group of people by ticking off boxes on a checklist of Hispanic heritage, which was precisely the kind of thinking Univision was trying to pivot away from. Jose Luis Revelo, Man Made Music's director of music strategy, Latin markets, became instrumental to the process. He had been born in Colombia. At age eight, he moved with his family to Panama. They lived in the area around the canal, so there were lots of Americans. "I've been a nomad all of my life," he says. "My entire upbringing has been bilingual and bicultural." Along the way he has developed a deep understanding of Latino and Caribbean music as well as a broad knowledge of all types of popular music. He knew the musical language and instruments we had to use to weave together a diverse cultural fabric for Univision's audience.

The nuance here was typified by clave, the five-stroke rhythm derived from sub-Saharan Africa. It's the root of modern Caribbean styles such as salsa, merengue, bachata, and reggaeton. It's the sound at the heart of Afro-Cuban music. One thing it's *not* is Mexican.

Mexico traces much of its musical heritage back to Europe by way of the conquistadors. When Mexican listeners hear a rhythmic pattern that usually belongs to the Caribbean, there's a huge risk they'll say, "That's not for me." Reggaeton sensation Daddy Yankee is an artist who typifies this idea. His song "Gasolina" was a huge crossover hit in the United States with both Spanish- and English-speaking audiences. But Daddy Yankee has never gone over as big in Mexico as he has in the Caribbean, despite efforts to market his music in Mexico. For Univision, a Daddy Yankee approach would have spelled an epic failure. We had to bring along the Mexican audience. More interesting for Univision's purposes was the Dominican artist Romeo Santos, who'd begun mixing Dominican bachata with Mexican ranchera with great success in both countries.

In one particularly heated discussion, we went back and thought about Linda Ong's insight that young Hispanic and Latino viewers wanted to watch the Kardashians along with their telenovelas. That's how we eventually realized that we shouldn't be searching for some magical concoction of Latin styles. We had to cut a vibrant cross-section of Anglo pop music and Latin pop music. We couldn't be thinking about modern artists with traditionally influenced styles, like Shakira or Gloria Estefan. We had to be as subtle and tactful as Rihanna or Pitbull. Non-Hispanic listeners might not even notice the Caribbean or Latin sonic signifiers in their songs. And Caribbean or Latin heritage isn't necessarily what defines those artists. But the clave rhythms are there in the background, adding to an experience that feels authentic to the Hispanic or Latino listener who might hear a bit of himself or herself in the songs. This is one of the secrets to massive-selling crossover pop artists who've found ways to capture the shared cultural values and unique emotions in music that sounds both authentic and new. Non-Hispanic artists are exploring this territory too. Rhythms in songs by 2013 chart-topping Seattle artists Macklemore and Ryan Lewis, for example "are unmistakably syncopated in a way that is definitely not Anglo," Jose says. "Of course, all of these things are debatable, but I think that rhythm owes more to the Latino background than, say, an African background." Our challenge for Univision was to include some of these elements, but in subtle ways.

At times, finding this balance came down to a single instrument. The accordion, for example, is tricky as hell. It's where the cultural legacies of Mexico and the Caribbean intersect. If we let our accordion carry the melody (the notes) in a way that sounded something like polka, we'd be speaking to Mexican contingencies with their European lineage. Used as more of a rhythm instrument (the beat), the same accordion would speak to Dominicans, Puerto Ricans, and people with Caribbean or African cultural heritage.

Early in the process, in one of the pieces, we included a snare drum that played in the kind of pattern you might hear in reggaeton. Somebody reacted to it in a meeting, saying it sounded "a little too Caribbean," Jose recalls.

The anthem had to speak to *everyone*.

Language would send signals too — and not just the meanings of the words but how they were pronounced. Just as some instruments have cultural connotations, so do some words. For example, some Caribbean people tend to drop the consonants at the ends of words and let the vowels hang. So *vámonos* (let's go) becomes *vámono*. "That is absolutely not Mexican," Jose says. One of the singers we'd hired had that regional, cultural thing going on. "His background being Puerto Rican, he would get in the heat of the moment and get inspired and say something that is typically Puerto Rican," Jose says. "I had to tell him, 'Let's not use that last one, because not everyone's going to get it — it's not part of the common Mexican colloquialism' . . . and there are *lists* of colloquialisms."

In the end, the anthem for Univision sounds like a celebration. Baked from a blend of traditions from both Mexican and Caribbean musical sensibilities (and, yes, with prominent electronic textures *and* an authentic accordion), it sounds like the vibrant cross-section of contemporary Latin and Anglo pop-music culture. Lyrically, it's a celebration of family and life, and it seems to resonate in a moment when Latino culture and growing political power has come to the fore in the United States, and when Hispanic Americans are getting their due as the growing middle class moving the country forward.

This is the tangible, concrete stuff of anthem-making. Even for Jose, who's steeped in history and the culture of Hispanic and Latin music, the Univision project was an education. "The Latino or Hispanic identity is a distinct thing," he says, "more than being Caribbean or being South American or being Spanish. And what's happening in America is that it's producing this very peculiar mix and intermingling of cultures—and its own demographic. I think we have to start thinking of Latinos in the U.S. as their own animal."

Univision did.

After a several-month development process with the Univision marketing team, the new anthem and its sonic logo premiered with all employees December 2012 and premiered across the entire network in news, sports, and promos on January 1, 2013, with the radio scheduled to follow later that spring.

Fast-forward to July 2013: For the first time, and for four weeks in a row, Univision won the prime-time ratings among eighteen- to thirty-four-year-old and eighteen- to forty-nine-year-old viewers. It beat second-place Fox by double digits.

There were plenty of reasons. The network's go-to programming, including sports events and a popular soap opera, played big roles. So did *Premios Juventud,* the youth awards show, on July 18, 2013, which averaged almost five million total viewers. But Ruth believes it's a sign that Univision is on the path to transforming the brand into a movement. In the thirty-six months leading up to late 2013, Univision Communications International had expanded from three networks to fourteen, plus a joint venture with ABC called Fusion, which takes direct aim at bilingual, bicultural young Hispanics and Latinos; one sign of this is that the news is in English. The company had gone from having no digital presence to launching a network of websites and YouTube channels.

Sound, Ruth says, is playing an increasingly important role.

"Randy was part of the team that did NBC's sonic branding," she says, referring to Univision's CEO. "This is not a new trick for Mr. Falco."

Our sounds — from the full anthem to the distilled logo we created — were starting to show up whenever and wherever Univision wanted to indicate its alignment with the new Hispanic movement. "With just three notes, you can signal you're part of this thing," Ruth says. "And when those notes happen, you have assurance that this is going to be quality programming and you can have assurance that the community's going to be behind you."

Ruth believes, as do I, that sound only makes good business sense as Univision expands and diversifies in every way. It's part of the reason you've probably started to hear more about Univision, even if you've yet to tune in. With sound, Ruth says, "You will increase brand awareness, you will increase eyeballs, you will increase ratings. And ratings you can take to the bank."

Strategic sound does more than just amplify an intangible feel-good idea. It gets results. And when a sonic identity syncs with the visual identity to tell a single story across every multimedia channel, the impact is exponential. A brand and its anthem become something more. They become part of a movement.

7

Scoring the Experience

N 2010, NBC HAD one hell of an NFL football season. It averaged 21.8 million viewers for *Sunday Night Football,* up 2.4 million from the previous year. It was the number-one show of the prime-time season for every single one of its eighteen games. No other network could claim such success. By all accounts, everyone in football-loving America was tuning in.

Then came the Super Bowl XLV broadcast on Fox on February 6, 2011. It averaged 111 million viewers, almost four times more than it got in the regular season. It became the most watched program in U.S. history.

It was not that there were 89.2 million football fans who watched only the Super Bowl and none of the other games. Actually, the surge in viewers was due to people who weren't avid football fans at all. "They're doing it because the rest of America's doing it, and they're caught up in the wave," says Fred Gaudelli, NBC's go-to producer for the Super Bowl and *Sunday Night Football.* The big game always draws way more watchers. That's because the biggest football game of the year is the one that has the least to do with football. The Super Bowl isn't for sports lovers. It's for everybody else. "You want to entertain those people," Gaudelli says, "and they're not going to be entertained by strategy and x's and o's. Because they're not going to understand it. And you're not going to teach it to them in one sitting."

The Super Bowl has to feel less like a sporting event and more like an epic Hollywood movie. And what's a movie without a score?

Music is one of the most powerful tools a network has when it comes to raising the game-day stakes. It can also help meet seemingly impossible business goals. Each year, the NFL charges the networks higher licensing fees for the game. In turn, the networks hike the rates for advertisers — the price tag for a thirty-second ad during Super Bowl XLVI was estimated to be four million dollars. And for such a steep rate, advertisers expected higher ratings than the previous year. Fox's 111 million viewers for Super Bowl XLV in 2011 beat CBS's 106.5 million viewers in 2010, which at the time had set a new record for the most watched program in U.S. history. Now it was NBC's turn.

Marketing and the promise of a spectacle — Janet Jackson, Justin Timberlake, the Black Eyed Peas, or Madonna at halftime — can help attract a portion of new viewers to the game. But sponsors expect that people will stick around for the entire event, which lasts twice as long as most movies. The most watched single play of Super Bowl XLV on Fox, according to TiVo Inc., was when Steelers' quarterback Ben Roethlisberger threw an incomplete pass to Mike Wallace. There was only about a minute left in the game.

Getting a broad range of viewers engaged until they see how the story ends is Fred Gaudelli's specialty. He's quick to share credit with his production team, but Fred's a big part of why an NFL broadcast on NBC feels omniscient — he's one of the guys who started putting cameras on backfield players to show how they affect what happens to the guy with the ball. And he gets the power of sound. "We always try to put people in the best seat in the house," he says. "But the best seat in our house comes with sound." That's why you hear the sounds of helmets crunching, the quarterback yelling signals, and bodies hitting the turf. Sounds bring the game to life. But music turns it into a story — it's the emotional engine.

You might know that the *Super Bowl on NBC* has its own theme. You might even be able to hum regular-season game-night music from Fox, NBC, or ESPN, but odds are you're not paying strict attention to all of the supporting pieces of music in those broadcasts. You might realize you're hearing versions of the *Super Bowl on NBC* theme pop up before commercials or as the show returns from a commercial break, but you probably aren't aware of how instrumental those bits of music are in keeping you engaged and transforming the Super Bowl into a universally relatable drama. You don't have to understand it to feel its power.

"Music is one of the highest forms of entertainment that I know," Fred says. Neuroscience and brain imaging back this up. Sukhbinder Kumar, a staff scientist with Newcastle University's Auditory Cognition Group whose interest is sound and emotion, has repeatedly found in studies that music is a powerful trigger in all kinds of experiences. "There's nothing like it that evokes such strong emotions," he says.

You might not even realize you're hearing them, but key moments of music are setting expectations and telling you what's at stake. They're establishing mini-cliffhangers and recapping and reinforcing important plot twists. That way, even viewers who never watch a single regular-season football game don't feel lost and still want to come back after those four-million-dollar commercial breaks. On a functional level, musical moments let you keep track of the drama even when you're not looking at the screen. The Super Bowl fanfare cuts through a room, finds you in the kitchen, and tells you: *Drop the nachos; the action's back.*

On a deeper level, music highlights and elevates the moments that build on the Super Bowl legacy, so you feel them as they happen. The right music at the right time takes a play you just saw and makes you feel like you've witnessed history. And it makes you feel all of this precisely when you need a reason to stay in the story. It's the same tool directors use to keep you on the edge of your seat during a film or to support key emotional moments that drive the most complex plots forward. "If you talk to any director, they'll say music is fifty percent of the movie," says the film composer Hans Zimmer. His score played a vital role in Christopher Nolan's *Inception,* a movie, coincidentally, about implanting ideas in people's minds. The 2010 film was nominated for seven Academy Awards and won for cinematography, visual effects, sound mixing, and sound editing, despite having an intricate plot that involved alternate realities, time-shifted dimensions, and dreams within dreams within dreams. "Music is how you can get away with a very abstract story liked that," Zimmer says. "It was a subliminal way to take you on that journey. It was emotionally comprehensible even if you somehow missed the odd line or intellectually you have a problem rolling with it, which is okay. Music just turns everything into an emotional experience."

On NBC, the Super Bowl becomes *Star Wars.* That's mostly because of its theme, "Wide Receiver," composed by John Williams, who scored *Star Wars, Jaws, Close Encounters,* and just about every other award-winning film that generations of people remember. You need to hear only a note or two to recognize it. As Fred Gaudelli put it, "Right off the bat, you're starting with a pretty heavy piece of music."

That's also where I come in. For the 2012 broadcast, NBC asked me and Man Made Music to give Williams's beloved score a number of overtly modern twists.

In 2012, the ad prices were higher and the ratings expectations were larger. NBC execs were looking to retain the DNA and the popularity of the original score but modernize the style and extend the theme. They had their regular-season music but needed more to draw from for the Super Bowl, which told a dramatically bigger story with an epic scope. Williams had written a beautiful and timeless theme. My job was to make it of the moment. The Super Bowl itself is timeless, but with every new Roman numeral, it builds on the collection of moments. It is a real-time unveiling of players on their way to becoming MVPs or even the greatest of all time. And it's not just their Hail Marys and game-changing marathon runs but all of the universal emotions that go along with those plays. Those moments had to instantly make sense to viewers who didn't know a noseguard from a tight end.

Emotionally moving sounds have to fit in three- to fifteen-second spaces during the broadcast. In those snippets, they have to spark feelings and memories you associate with years' worth of Super Bowl experiences — not just what happened onscreen but the time of year, the lingering winter chill in the air, the friends gathered around a single TV, the celebrations before and after the game. Sound like a tall order? It's not for music, even in those short-form bursts. Cognitive psychologist and Western Washington professor Ira Hyman has published research on this particular power of music, memory, and emotion, and he has written regularly on the topic for *Psychology Today*.

"Songs that are distinctively associated with a time period or a series of events . . . can act as wonderful memory cues to both bring to mind those memories and then to also bring to mind not only the emotion experience of the time but that sense of nostalgia," Hyman says, citing studies that have found positive correlation between emotion and memory.

Not only is music an emotional engine, it's a Magnum V-8 that gets mileage like an EV.

But for the *Super Bowl on NBC*, we couldn't just come up with more music. My team and I had to tap into the right styles of music that would acknowledge a new generation of watchers. The new styles would build on a sonic vocabulary. And it couldn't sound like pandering. We had to connect with new viewers and be relevant in that cultural moment, meaning we had to infuse a whole range of musical styles into John Williams's original composition and extend it with rock guitars, hip-hop drums, electro beats, and, yes, dubstep. But most important, we had to be true to the work and evoke a wide variety of emotions to cover all the anticipated story points of the game: tension, triumph, optimism, driving energy, and others. It was like scoring a movie before it's been filmed.

On top of everything, our music would have to cut through the roar of a crowd and sync with the narration of commentators Al Michaels and Cris Collinsworth. And we had relatively few chances in a mere two-hour, thirteen-minute, forty-four-second broadcast (not including commercials) to drive forward an entire spectrum of emotions. Oh, and as is often the case in my business, the deadline was insane. TV moves quicker than most businesses, especially when it comes to production and sound. And live TV scrambles like an exposed quarterback in a full blitz. We got the Super Bowl assignment precisely three weeks before kickoff, effectively meaning I had two weeks to write the music, book the orchestra and various players, record their parts, mix them, arrange them, and submit them to NBC with time for producers to rehearse.

Our variations on Williams's themes would become tools that Fred and his senior sound engineer would use during the game as events unfolded. Neither of them knew which ones they'd need or when, exactly, but they knew they needed them ready to go by game day. They'd categorize the sounds we provided by their emotional component and set them up so they could be triggered in an instant. "The script is unfolding live," Fred says. "We're seeing it the same time you are."

Super Bowl XLVI marked the second time NBC had come to me with the assignment. In 2008, my team and I recorded eight new derivative works based on "Wide Receiver." In 2012, we recorded four more to address new musical styles that had become popular in the course of four years. Throughout the process, we had to keep in mind that we weren't working on merely the Super Bowl. We were working on the *Super Bowl on NBC*. No one network owns the entire history of the Super Bowl. They take turns licensing the broadcast rights and the chance to put their stamps on a piece of history. Even without watching too closely, you can see the hallmarks of the networks, their different attitudes. Fox is all about over-the-top, often robotic-looking superhuman gladiators. CBS is a bit more about the legacy. NBC is about the epic celebration of sports. Each network establishes its own identity over the course of the season, and they all bring that

DNA to the Super Bowl. If the Super Bowl itself is Disneyland, the networks are like the worlds — Frontierland, Tomorrowland, Fantasyland.

I'd gone through that part of the drill with John Williams before. I had the honor of working with the master film composer in 2004 on a twentieth-anniversary update of "The Mission," the iconic piece he wrote for NBC News in 1984 — two of its four movements are still used on *The Today Show, Meet the Press,* and *NBC Nightly News.* The original score held up remarkably over time, but changes in broadcast styles and the way stories are told had meant the network needed a significantly different approach to creating shorter elements and extending the melody for different cuts and edits of the original theme. We recorded and produced a one-hundred-piece orchestra (the largest orchestra on television) at the legendary Sony Pictures Studios in Culver City, California, the same place Williams had recorded most of his timeless scores and soundtracks. I'd sat in the sound booth — the rest of the orchestra was set up in the giant room beyond the glass — with Williams and a trumpet soloist and watched the composer push an already powerful player beyond what he thought were his expressive limits. (Williams music is notoriously tough on horn players — plus the Culver City studios diminish the relative power of strings, putting even more pressure on the horn sections.)

What I learned, among other things, is that Williams is remarkably adept at *translation.* Not only does he internalize the storytelling demands of the client (a network or film director, for example), but he processes the language to determine the underlying emotion. Then he translates that emotion into the language of an orchestra — a "breaking news" vibe would, for example, become agitato (a musical term that means "restless"). Williams would know exactly which parts needed to soar to make a piece rise and meet the emotional demands. During the Culver City session, for the theme that was to play beneath Tom Brokaw's specials, Williams singled out the first trumpet, a player named Tim Morrison whom he'd worked with in

Boston during his time with the Pops and convinced to move out to Los Angeles. Williams brought Morrison into the sound booth to record his part rather than keeping him out in the main room with the rest of the ensemble. Williams was pushing him to play a personal best. And Morrison delivered when he performed his solo—his notes, pitch, timbre, clarity, and control were all breathtaking. When he was done, Williams reminded him how many people would be hearing that solo performance for another twenty years or more. "Once they hear this, every trumpeter in Los Angeles is going to be eating his heart out," Williams said.

That previous in-studio experience with Williams allowed me to summon the courage to tinker with the maestro's work for the Super Bowl project. But, of course, I deferred to him, asking before I recorded a note of my own if he'd want to review the scores and advance mockups. He's the preeminent modern composer for orchestra, a living legend. But John's agent got back to me and said, "No, actually, Joel, John said you should just do what you think is right."

That scared the living hell out of me.

The best way to explain the sort of power we were wielding when it came to the Super Bowl is for me to take you back to the way Disney deftly handles the power of sound. At its theme parks, you're overwhelmed with bright colors and cartoon characters come to life, but sound is still the most powerful emotional engine. The sound at Disney parks never stops working. Almost everyone who's set foot in a Disney theme park has felt the famed magic. Sound is at least half of that magic. It hits you in the parking lot and goes with you when you leave. Throughout the experience, Disney uses sound to usher you in and out of little dramas—rides, worlds, and shows—in much the same way NBC uses sound to transition you in and out of commercial breaks without ever letting you forget to come back to the Super Bowl. Disney uses ambient music even in its attraction lines. When rides or shows are over, sound tactfully prods you along. Whether or not you're conscious of all of these Disney sounds, you feel them.

They constantly pull you through the fantasy. Disney's grip on its sonic environment is so tight, they even score the bathrooms.

"Once you get inside our berm, you're in our world, and we want you to stay in our world," says Disney Imagineering's principal media designer Joe Herrington. Every square inch of the park is intended to be a magic place apart from the real world.

The show also starts long before you might imagine: right after you leave your car. Disney painstakingly crafts a script for tram operators to recite as they're whisking you through the parking lot. Operators are coached on proper microphone etiquette and use. If someone sounds like he's having a bad day, that can blow the experience right from the beginning, Herrington says. "Most of us [who work for Disney] will in fact go to that ride operator and say, 'Get your act together!'" The same goes for operators who put the mikes too close to or too far away from their mouths. If you've ever ridden a New York City subway and tried to understand what a tired MTA conductor was saying over the PA system, you know what this is like — if a mike is too close, the sound is garbled, distorted, loud, and fuzzed out; too far, and it's a distant whisper or mumble, maddeningly indistinct at the exact moment you most need clear direction about where you're going or why you're not there yet.

When you hear that crisp, informative voice on the Disney tram, you start to get a sense that something here is different, and not just because of the words the operator is using. This isn't any old amusement park. The streets are clean. The seats are clean. But most of all, the sound is immaculate. By the time you approach the ticket booth and come through what Herrington refers to as that "early development zone," you start to feel full-on Disney. "You go through those gates and you're inside, and all of the sudden you've got that foreground music that says, 'You are here in this magic place,'" Herrington says. "And from that point on, every place you go, there is a story being told to you on a subconscious level."

Everyone knows "It's a Small World." Most of us can hum "Yo Ho (a Pirate's Life for Me)" from the Pirates of the Caribbean ride. But there's a world of subtle sounds playing almost everywhere at Disney theme parks. Speakers are designed into the landscape. They're tuned and positioned precisely to hit you when you need sound the most.

"Half of the storytelling ability is sound," Herrington says, echoing what Fred Gaudelli, Hans Zimmer, and just about every savvy person in the storytelling business believes. It's an emotional place-setter that helps you feel precisely the right way at precisely the right time, Herrington says.

At Disney, the sound often kicks off the experience with lots of what you subsequently see or interact with. It works in perfect harmony with costume colors and set decorations. "If you just had a person walking around in a costume, you could get that at Chuck E. Cheese's. But you have a person who's in costume there against a bed of flowers that are perfect. And then you have the music that is supposed to tie those things together to create a sense of place," Herrington says.

Disney parks are divided into lands, and sound and music are crafted to tell you where you are before you see a single cowboy or space man. So on Main Street, U.S.A., for example, at Disneyland in Anaheim, California, you might hear "Meet Me in St. Louis," "Alexander's Ragtime Band," the tinkly piano runs of Scott Joplin's "Maple Leaf Rag," or even Michael Giacchino's "Married Life" from Disney Pixar film *Up*. You might hear the dinging bell on a replica of an old-timey fire engine or, if you're in Disney World in Orlando, Florida, the chugging sounds of the Walt Disney World Railroad. The vibe is heritage and familiarity, and you don't have to see the white-columned buildings to feel it right away.

Hike down the litter-free road just a bit, and you'll hear Adventureland. You might hear (even if you don't notice) inviting music beckoning you toward an event or an experience. You'll find the iconic Pirates of the Caribbean ride. Herrington says Pirates of the Caribbean is always on guests' to-do lists because it's more than a ride; it's a story.

Along the path of the Pirates boat ride, songs and sounds trigger you to turn your head and dive into a new scene — the pirates enjoying the spoils of a recent plundering or other salty dogs sitting out sentences in crusty jail cells. Behind you, as you turn your attention away, the robotic scene resets for the next group of guests moving through the attraction. Each time, sound and music set that stage. It comes first by design.

"We are a very visually oriented society. But you take sound away and a lot of visuals are going to be misunderstood," Herrington says. "Visuals can't set a mood as quickly as sound can."

When the ride ends, you exit and it's over, right? Wrong. At Disney, sound, story, fantasy, and magic never stop. And sound is how the company moves crowds at the parks. It tells you, *Leave!* on a subconscious level (and less rudely). You know what exit music in a situation like this sounds like. Tempo is a big part of it, Herrington says — the energy and tempo and the way an upbeat number might contrast with everything you've heard up until that point. Or musically, a piece might obviously resolve, telling you, *Okay, it's over. You can leave without missing anything.* Then comes the march. "You can't sit still when you're listening to a march," Herrington says.

Still, sound is far from done with you. The job of sound at Disney Parks is especially important when you're least likely to be aware of it: between rides or lands. After all, when you leave Pirates of the Caribbean, you haven't left Disney. And you have to feel the place at all times, even as you're coming down off an emotional (or actual) roller coaster. The areas between rides or theaters or lands at Disney are places where guests decompress. For those spaces, Herrington and his team might create natural ambient textures or include subtle music that gives you permission to sit down and relax without the fear of missing out.

At Disney, these are called transition and decompression zones. They're how you mentally detach from the world you just experienced while staying stitched into the overall Disney fantasy. It's the glue in your total immersion, that state that feels like magic. These areas are ushering you out of your last experience, but they're also helping to prepare you for the next one.

While you might not be aware of these sonic buffers, they're actually tailored to the way your brain works. The sound and music in every land is created to tell a story. If they overlapped, they'd be fighting for your attention. "The minute a guest walks through a zone like that, confusion sets in," Herrington says. "They don't know what they're supposed to be listening to. They don't recognize it as being wrong, necessarily. It's their subconscious that says, 'I'm confused, I don't know what I'm supposed to be paying attention to.' And then they're out of the story."

The quickest way to appreciate these buffer zones and the role they play in Disney's magic is to imagine what it would be like if they weren't there. Without them, you might feel like the story's escaped you.

The catch, of course, is that not only does all of this have to happen in the span of seconds, it has to happen invisibly. "If sound is in your face and in the way, we are not doing our job," Herrington says. "We're doing our job when you don't really notice it, because it just feels correct."

A lot of my job for Super Bowl XLVI on NBC in 2012 was creating some new notes and musical areas of development. Williams had scored the attraction. I was there to help usher our viewers off each ride, help them reset, and bring them back for another thrill. I couldn't keep an ambient version of "Wide Receiver" playing beneath ads for GoDaddy.com or Audi, but I could send viewers off from our story and into those spots with a reminder of what dramatic moments had just unfolded and what would be at stake when they returned. My team and I gave Fred and his team almost seventy snippets of music in virtually every length, style, and mood we could anticipate — in time for him to sit down and map out which types of sonic themes worked best for which types of situations.

During the game itself, you probably felt elevated levels of emotion after key moments. Whether you realized it or not, music had a lot to do with that. Here's how, sonically, it all went down.

In the run-up to the game's kickoff, announcer Al Michaels called the Giants-Patriots game "the sequel, the rematch, the encore, act two, back to the future" (referring to Super Bowl XLII in 2008, when the Giants beat the Patriots). It was an anxious opening moment that defined the stakes, the kind that might have made you suck air through your clenched teeth. Fred Gaudelli's team played a cut we called "Play Action."

As Patriots' wide receiver Rob Gronkowski warmed up and struggled through a high ankle sprain, viewers could almost feel the pain themselves. Part of that spiky tension was carried by a piece of the reworked Williams theme called "Epic Matchup."

To kick the high tension up yet another notch when the Giants trailed the Patriots by less than a touchdown, Fred's production team played a sped-up, guitar-laden, electro-beat theme called "End Zone E."

When he wanted to trigger feelings of triumph after the Giants seized momentum, Fred's senior sound engineer played a trumpet-heavy, chopped-and-remixed version of Williams's original score that we called "Backfield."

Music even helped you feel like you were speeding through NBC's need to pay the bills: a fast-paced cut of Williams music played behind a sponsor roll call — Bud Light, Audi, Pepsi, Best Buy, and more.

But the best example of the score's role in the big game came during the final plays of Super Bowl XLVI. Remember, this game had been hyped for two weeks. "So by the time it comes, people are frothing," Gaudelli says. "Now you gotta pray that a really good game breaks. You can try to tell good stories that are interesting. But you want a cliffhanger going down to the final play, which is what 2012 was."

Things got really intense just before the two-minute warning. Unique to the NFL, this break in action was originally developed to let the official in charge of timekeeping sync his clock with the play clock on the field. That changed in the 1960s, when the stadium clock became the official timekeeping device, but TV had become important by then too. The two-minute warning became a major moneymaker for the networks. The inescapable fact is that, at the most intense moment of the broadcast, NBC was going to commercial, and couch jockeys everywhere would have to chew on their nachos and nails during five minutes' worth of sales pitches.

As Super Bowl XLVI transitioned from the two-minute warning into commercial break, a graphic flashed on the screen: "Final Act." And Al Michaels set the stage for suspense: "Big finish ahead. It's Super Bowl Forty-Six! New England . . . by *twoooooo* . . ."

Most viewers felt suspense, worry, anxiety, anticipation, disappointment, disbelief, surprise, or pure joy. This is precisely where music carried the most weight — and created one of the most important Super Bowl boom moments. Sound prolonged the nation's exhale and probably made a few sports fans remember their first trips to ball games. Along the way, the score was how NBC put its stamp on a broadcast with a massive viewership, part of a decades-long legacy. Imagine if no music played in those twelve seconds. Fans would feel tossed off into a sea of ads with no life preservers. If the wrong music played, they might feel ejected from the story. In Disney terms, they'd be going from Frontierland straight into Tomorrowland with no sonic transition, no area of development.

But the right music did play, thanks to Fred and his team. And a visual collage of intense moments flashed across the screen. Viewers saw slo-mo shots of Giants wide receiver Mario Manningham's now historic catch, both coaches' joy and frustration and grimaces and grins. There were spikes and celebrations and beefy linemen getting on one knee and pointing toward the goal. There were cuts to shots of Giants quarterback Eli Manning, Patriots QB Tom Brady, Giants defensive end Jason Pierre-Paul, and others. Everything about those twelve seconds signaled an epic matchup, a replay of the two teams' 2008 Super Bowl XLII game, which the Giants won, 17-14. More and more, this was looking like a déjà vu. A do-over. And in the end, it was almost exactly that.

After the two-minute warning and a smattering of nerve-racking plays, with five seconds left in the game, Brady was down to one chance. From the forty-one-yard line, he fired a bomb into the end zone. Six Giants swarmed on two intended Patriots receivers. The ball was tipped, and the Patriots' Gronkowski dove for it, but the ball hit the turf inches beyond his reach. Game over. Giants won. "Epic Matchup" faded up slightly in the background, this time the full version. The game had ended, but every good drama needs a denouement.

The Giants won the Vince Lombardi trophy. Eli Manning, having secured his reputation of being one of the greatest clutch quarterbacks in history, accepted the MVP trophy. "Epic Matchup" would bookend the broadcast. As the highlights of the game played one last time, viewers heard the swell of heavy-metal guitar chords we'd added to the new arrangement of Williams's theme. The scene cut from the winning moment in the game to a montage of the reactions afterward. Everything went slo-mo. Eli Manning jumped in the air, hugged his teammates, and then went jogging toward the middle of the field; high fives erupted; red, white, and blue ticker tape streamed down on Manning, who was at the center of a media and security scrum. Cut to Tom Brady, whose head lowered as he unsnapped his helmet chinstrap and trudged across the field. The final minutes felt like the end of a movie. The sound put order and emotion behind the chaotic scenes of the Super Bowl. It transformed them into a farewell. A crescendo of horns and guitars and drums and strings swelled one more time and built tension. The ticker tape glittered once again. The Gatorade flowed over the head of Giants' coach Tom Coughlin. Al Michaels and Cris Collinsworth had gone quiet. The final *Super Bowl XLVI on NBC* logo appeared and took over the screen one last time, and the experience ended. The score resolved, but you didn't have to know a thing about music to feel it.

Short of hooking up electrodes to Nielsen families, there's no way to scientifically measure whether music is working. It's an overall feel that draws upon a lot of factors. To torture the football metaphor, sound is like the final four-yard drive through a crowded noisy defensive line, the push that leads to the touchdown. It's not a long pass. It's not the play of the game. It's an effective run, often without heroics. If I did my job right, I absolutely *didn't* expect to hear the music called out specifically. Sound wasn't solely responsible for the amazing fourth quarter of Super Bowl XLVI. But it elevated the drama, even when the drama was plenty high on its own. Plus, sound is never neutral. If it distracted you from what you wanted to feel when you wanted to feel it, if it got in the way of the intense action, you'd know. If it sucked, you'd remember.

Do you?

Super Bowl XLVI broke Fox's 2011 viewership record to become the most watched television event in U.S. history, averaging 111.3 million viewers. The peak ratings happened way beyond the reach of the advance buzz — three and a half hours in, between 9:30 and 9:58 p.m. (the last half hour of the game) — when an average 117.7 million people were watching, according to Nielsen.

I'd scrambled to write everything in a week, since I had only a week after that to bring in the orchestra and musicians and record everything and finish in time for NBC to rehearse for another week. With Williams's blessing, we'd done ten different arrangements blown into seventy cuts for every imaginable emotion. And per Williams's request, when it was all done, I had sent him a CD with all of the different cuts. Two weeks went by and I didn't hear a peep from his camp. All I could think was, *Oh shit.* I knew Williams had seen the game, because he's a huge football fan. But I also knew him to be a private guy. Finally, I couldn't take the suspense any longer. I called up his agent and said, "Just curious whether John had a chance to review all of the music."

"Oh, let me find out," he said. Then he hung up.

It took until later that day to hear back. When you're dying to hear something, silence can make the wait feel like an eternity, another powerful way sound works. Finally John's agent rang me back and said, "Yeah, John said, 'Good job.' He was pleased." And I was relieved.

8

Creating Boom Moments Every Day

Y OU CAN'T HOLD SOUND, put it in a basket, weigh it, or admire how it looks. You can't take it for a drive, touch it, or inherit it from a rich relative. It's just energy in space. But you *can* actively be aware of it and apply it to great advantage. You don't need to be a musician, an expert, a DJ, or a great orator either.

Remember Dr. Sean Olive, Harman's director of acoustic research? In his study, he was trying to figure out whether people cared about sound quality in headphones and other audio gadgets. He tested an array of headphones and asked for sound-quality ratings from two sets of listeners: trained and untrained listeners. Surprisingly, he learned about as much from trained listeners as he did from the untrained listeners in this study. "Untrained listeners tend to like everything a little more," Dr. Olive says, but both groups rated sound on about the same curve. They showed about the same ability to tell good sound from bad. In the end, regular people "have the innate ability to make quality decisions, especially when they're put into an environment where everything is controlled," according to Dr. Olive.

You already have the ability to be discerning about sound. You don't have to be one of Harman's trained professional listeners to get more out of sound and to create boom moments in your everyday life.

In this book, I've touched on the neuroscience, psychology, and sociology behind a bunch of boom moments. You've seen how sounds, even short sounds, can convey information and emotion and even help you determine how to act. I've shown you the principles of effective sonic branding and told you stories about how they've played out for some high-stakes business challenges. This chapter is about you. Boom moments and opportunities for boom moments are everywhere. Every day, you encounter countless opportunities to take advantage of them or make them yourself. There's no way to predict every scenario in which sound could affect your life. But in this chapter, you'll see how to use sound to transform all kinds of situations. Once you begin to consciously think about soundscaping, that practice of fine-tuning your awareness of these sounds, you can make more of the subconscious conscious. And you can design your own sonic experience — for yourself or others.

Boom moments are emotionally compelling sound-based solutions to everyday problems. They can transform potentially embarrassing situations into confidence boosters or add meaning to mundane activities. Simply put, they enrich our lives and give us critical information to help us navigate our world and get what we want. They also help us create profound moments of experience where we feel most alive and engaged. Boom moments are moments of connection to ourselves and others. They are part of what makes us human.

Whether you realize it or not, you begin crafting your personal soundscape the minute you wake up. Think about the sound of your alarm clock. Is it one of the blaring preset rings on your mobile device? Or do you have a more Zen alarm clock that wakes you with a chime borrowed from a Buddhist monastery? No one likes to wake up to something blaring. It's startling. Or at the very least, it produces a negative reaction. Sometimes people even anthropomorphize whatever is causing the harsh waking sounds and think of it as if it were a rude human rather than a device. "All right, all right!" you might bark at your inanimate alarm. Conversely, you might sit and stretch and enjoy a softer waking tone or get information from an alarm that begins softly informing you of the news. (Some smartphones that run on the Android operating system have a function that begins relating the day's events, headlines, and appointments as you wake.)

Now that you're awake, think about what your home sounds like. I've shown you the principles behind the effective use of sound for retail spaces. But much like the air you breathe, the world you see, and the food you eat, the sound and music around you shape your personal environment.

Even before you take your first step, the volume of birds chirping and singing (or of traffic honking) outside can instantly clue you in to how early or late it is. Something as simple as a refrigerator running or the hiss of a radiator can help you feel a familiar awareness of home long before all of your senses have come all the way online. Most of us don't actually need a fireplace for heat, but the crackle and hiss of the burning wood can cue up warm memories if you stoke the flames on a cold winter morning.

Most homes aren't crafted with much consideration of sound, though. For the most part, our attempts to mitigate sound are rooted in what *looks* right. We pay attention to the visual layout and materials used in our homes based on what looks right too. We labor over the color or texture of tiles, wallpaper, or flooring, but we rarely consider how those materials interact with sound. We don't think too deeply about things like the angles of walls. Mostly we build ninety-degree corners and parallel walls, which can cause inside sounds to bounce around and create a perfect storm of progressively building noise in the middle of a living room (which explains why parties can get so loud).

In urban environments, where many live vertically, people often accept that city noise or footsteps above their heads are just part of city life. They wear earplugs or gulp down drugs to help them sleep. But some of this plague of noise could be solved with different sounds — running water, wind, white noise. You might not think you have much control over invasive sounds coming from outside, but low-cost noise machines can help mask unwanted sound and create the effect of silence. In New York City, for example, turning on a fan can drown out loud revelers or car horns outside your window or a lead-footed neighbor above you — or both!

Since you now know that silence is really about negative space, you can curate silence — or the experience of silence — in your home. It can frame the sound and allow space for the energy to occur — just like it does at Disney.

As much as we might think about insulating our living quarters from external sounds, we don't put the same level of importance on the way our homes interact with the sounds we actually *want* to hear. A cathedral ceiling is beautiful, but what if it prevents you from being able to make out the voices coming from your television? A wall-less second-floor loft bedroom creates an open, expansive visual appeal, but what does that matter if people in the living room below can hear your every whisper?

Think about this as you walk through your home or apartment tomorrow morning. You might have spent a lot of effort training yourself to tolerate noise or sonic trash in the place that matters most to your daily life. But even your home's sounds aren't the sounds closest to you on any given day.

As you get dressed, take a moment to think about what you broadcast through the sound of your apparel. We're all well aware of the visual impact of fashion, but how many people actually *listen* to an article of clothing in a dressing room before taking it home, putting it on, and going out? How embarrassing is it to realize there's a rhythmic *vvvvip!* with every step in a new pair of pants?

Not every fashion message is wrong. The sound of your clothing can convey useful information to others. When you hear the sounds of a nylon or rubberized raincoat or of wet rubber soles on a concrete floor down the hall from your windowless office, you know you should take an umbrella the next time you go outside.

The sound of clothing also conveys more subtle messages, influencing your perception of what people are actually like. What do you think when you hear jingling or clanging jewelry on a woman? How about on a man? Think about the story the sound of stretching leather tells you about the man or woman wearing it. Now think about that same sound in the workplace. Do you think of work boots and associate them with manual labor? Authority? How about a cop's utility belt? Think about what jingling keys on a ring tell you about the position held by the person wearing them as he or she walks down the hall of your office. What if it sounds like a whole bunch of keys jingling there on that ring? Is it the building manager or the custodian? Did you just realize you've worked so late that the cleaning staff has already arrived?

Nothing tells a story like a pair of shoes. Not only does the shoe itself do a lot of talking, but the shoe's sound tells you about a person's gait and even his or her personality. If you work in an office, whom can you instantly identify, without even looking up, by the sound of his or her shoes or walk? Is it the shuffle of an intern destined for low-level obscurity or the crisp clomp of an executive anxiously approaching you with a problem that needs solving, stat?

Are your own shoes sending the message you want them to? Ask a few people, and they'll tell you they feel more dressed up, more professional, or fancier when their shoes make a satisfying *click* or *clack* when they hit the ground. A hard sole means business and authority to most. There's a responsibility to live up to the level of authority you broadcast with hard-soled shoes or high heels — if you're trying to tell a lie with those sounds, you're begging to get busted, and then the sound becomes a reminder of your lie. But what if the sound those shoes made was actually interpreted completely differently by many of the people who hear it? If we're talking about a client meeting where you're trying to win business, or a job interview when you're trying to win a paycheck, this becomes a big deal.

"I would suggest that sound on a shoe that's marching is *not* appealing if you make a lot of noise," says Pasquale Fabrizio, the LA cobbler considered by many notable sources to be the best. To him, a loud shoe screams *cheap.* The makers of inexpensive shoes cut down on the cost of materials by making heels hollow. "If those heels are hollow, they're loud, and they're brutal," Pasquale says.

Being a popular LA cobbler means Fabrizio works with a lot of famous feet. A common task, he says, is making sure certain red-carpet-walking stars' shoes don't, let's say, *suck.* An actress whose name he declines to share came in once to have him silence the sound of a high-heel shoe. "It has an air pocket," she said. She didn't elaborate on the specifics of the offending suction-type sound; she only chuckled. "An air pocket?" he asked. What she didn't want to say was that the shoe she was planning to wear while walking the red carpet made a *fart sound.* Imagine the headlines in *Us Weekly.*

Then there's the message the sound of shoes sends about a person's gait and all it means. Pasquale suggests that a loud march is as desperate as a lazy-sounding shuffle. "It's like chewing gum," Pasquale says. "I don't know how many people say you could walk into a job interview chewing gum and think that's fine. The same thing applies with footwear. If you make sounds, it becomes annoying. Your shoes hit the ground and they're like, 'Look at me!'"

The devices you carry with you are also broadcasting messages, to you and to those around you. The world might be less annoying and noisy if more people thought strategically about the sounds coming from their phones. Here's a small but effective exercise in soundscaping you can do today. Take a few minutes to play with the sounds on your smartphone. Set up a specific, meaningful ring for regular contacts or loved ones and have it play when they call, text, DM, or chat you via Google Hangout. Notice how much time and effort or anxiety you save the next time one of them pings you.

Your phone alerts you to all kinds of stories. The right type of meaningful sound at the right time can provide exponentially more information around that story than a snippet of a pop song, the tune and tone of which probably don't match the incoming message. (Is it really appropriate to play Buckcherry's "Crazy Bitch" when your mom's calling?) And preloaded ringtones are often little more than brand sounders with no emotion or information attached. Without a word of text, the right strategic sound can tell you a loved one arrived somewhere safely; it can reassure you that communications or transactions are safe and secure; and it can create urgency around a work message or breaking news.

When you walk out the door to head to work or another destination, the symphony really begins. To those commuting in an urban environment, the cacophony can spark a quickening of pace. Voices mix with cars, wind, birds, and the sound of all of that technology. Instead of sounds hitting you one or two at a time, now you're trying to cut through the clutter to find your sonic lifeline — the approaching train or bus, your car's alarm or engine or radio coming on to signify the next chapter in your commute.

When you get to work, sound plays an even bigger role, enhancing or inhibiting creativity and productivity.

In a 2012 experiment, a team of researchers created a blend of sounds — talking in a cafeteria, roadside traffic, and distant construction noise — then altered the volume for experiments testing the impact of ambient noise. In a paper titled "Is Noise Always Bad?," the researchers noted that moderate ambient noise enhances the performance of creative tasks, promoting abstract processing in the brain, while a high level of noise is detrimental to creativity.

We've all felt this on some level; you can test out the principle and transform just about any space using a handful of productivity apps or services built on the idea behind this research. The Thunderspace app creates a bed of white noise intended to keep you focused. Coffitivity is a website and an app that mimics the sounds of a coffee shop at various times of day. Ambiance, another app, offers a wide array of productivity noises. New apps seem to crop up each month. These and others are easy to find in any app store.

Next time you're on a long flight or train ride, try to notice how much easier it is to focus on a task. The white noise does more than make your food taste blander and your snacks seem crunchier; it also masks distractions and lets you get busy.

I wrote some portions of this book with headphones in my ears and a low level of white noise playing so I had some negative space or silence even when I was in a crowded coffee shop, on a commuter train, or in a house full of noisy kids. White noise increases not only apparent silence but also the perceived distance between the space you're in and another sound source.

Once you've become adept in the art of harnessing sound to transform a space or to encourage productivity, you might start to think about the other ways sound can boost your performance and learning abilities.

Think about the way a lot of people learned how a bill becomes a law. "I'm Just a Bill" is more than a song; it's the ultimate civics lesson. It's almost impossible to think about spelling *Mississippi* without transforming the *i*'s into the downbeats of a little cadence. Phonics are a proven, powerful way students learn to read. In academic environments or when cramming to learn a language for, say, a trip abroad, seek out study methods that incorporate sound and music. Lessons are picked up quicker and last longer.

Sound can have a big impact on the way you are perceived by others at your work.

My colleagues and I often talk about how every brand and company needs its own distinctive sonic (or musical) voice. We're constantly thinking about what each one's story sounds like and what we can do with music and sound to let the listener understand and feel that story instantly. What makes each business tick? Where is it going? Where has it been? What's its identity, its values? What's the story behind the story?

When using music to answer these questions, you have to speak the language. I don't mean you have to sight-read sheet music. But you do have to understand the emotion, the story, and the experience that musical instruments typically add up to. When charged with the daunting task of expanding maestro John Williams's timeless Super Bowl theme "Wide Receiver," I had to introduce his anthem to a gamut of new instruments, part of the sonic palette I gave producers to help them score an unpredictable drama as it unfolded in real time. The right music was an important part of how they kept viewers glued to the experience, even during commercial breaks.

As an individual, you don't need a legendary composer or the voice of a celebrity or a classic rock band to tell your story with sound. And you can tell it *everywhere*. You already know how to broadcast your personality with your hairstyle, your car, the books you read, and even the music you listen to. You tailor all of those for all kinds of situations (a tux for formal situations, a limo for an impressive arrival, picking up a copy of the *New Yorker* — as opposed to *Maxim* — when an attractive woman is sitting across from you in the waiting room). But sound can also make a powerful first impression.

Think about your own personal soundtrack — the music you typically listen to on your commute, the tunes that help you face the day or that lift your spirits when you're suffering from too little sleep and too much stress. Now share a selection from that soundtrack at your next business meeting when you want to put your own stamp on it — with sound. Think of it as your anthem. Hit Play before colleagues arrive. Dial up an acoustic-guitar jam or pop if it's open excitement or positivity you mean to convey. Spin R&B or earnest classic rock if it's a more meaningful vibe you're looking for. Yeah, you'll get a few comments about using music this way ahead of a business presentation. Maybe a chuckle. What's better, though, a smile or a yawn? To be clear, "Don't Worry, Be Happy" can't make a spreadsheet full of quarterly losses seem like sunshine and rainbows. And you don't want to play aggressive rap, metal, or hard-thumping electro for anything, but the right, subtle song that matches the mood of your message can make that message resonate with your audience long after the meeting ends. It can instantly tell the emotional truth of a story before the first word is spoken. Then the presentation can confirm what your audience is already feeling.

The sound of your voice also has a huge impact on the way you are perceived. People use the sounds of the voices to make all kinds of snap decisions. *Is this woman serious? Is he for real? He doesn't* sound *like a cop. . . .*

In fact, the sounds you make are more important than the words themselves. "When vocal information contradicts verbal, vocal wins out," wrote renowned verbal-and-nonverbal-communication expert and UCLA professor emeritus of psychology Albert Mehrabian in his pioneering paper *Communication Without Words.* "If someone calls you 'honey' in a nasty tone of voice, you are likely to feel disliked; it is also possible to say 'I hate you' in a way that conveys exactly the opposite feeling." George Mahl of Yale University (whom Mehrabian cites) found that the *ums* and *ahs* people inject in their speech broadcast their discomfort or anxiety. Mehrabian worked out a formula from his research and found that the total impact of spoken messages is 7 percent verbal, 38 percent vocal, and 55 percent facial. Versions of this formula have been adapted for all sorts of everyday applications.

Consider Sarah Garrigan.* She was a business-school whiz kid who couldn't land a job, despite her saying all the right things in interviews. She discovered that it was the way she said them that was holding her back. Sarah had earned dual undergraduate degrees from Berkeley, in economics and business. She'd jumped right into work as a management consultant with one of the top three consulting firms in the United States. And just to round things out, she volunteered on the side as a financial counselor for single parents at a nonprofit aimed at alleviating poverty. Then she'd decided to get her MBA from Northwestern University's Kellogg School of Man-

---

* This is not her real name. While the gist of the story of the woman we're calling Sarah is accurate, some details have been obscured by Quantified Impressions to protect her privacy.

agement, ranked one of the top business schools in the world by multiple prominent publications.

In the first year of her two-year program, Sarah began the process of finding a summer internship. It's a stressful experience for any business-school student — competition is fierce. Top performers in the best internships typically get offered full-time jobs. Sarah was hoping to become a product manager for a consumer technology company, and she sent out more than fifteen internship applications.

Sarah's previous real-world experience helped her know just how to dress and conduct herself during interviews. She met with four Fortune 500 companies in the first week. None of them called her back.

Sarah was dumbfounded. Was it something she'd said? She thought she had made all the relevant points, expressed a high level of interest, and demonstrated that she had done her research and was knowledgeable and eager. She replayed the interviews dozens of times in her head before she finally arrived at an epiphany: maybe it wasn't what she'd said, but the sound of her voice.

That's when she found Quantified Impressions. The Austin, Texas, firm specializes in personal, professional, and organizational analytics for people like Sarah. It's developed a behavioral-assessment platform with the Kellogg School of Management, but lots of the inspiration comes from Mehrabian's work.

In a study, Quantified Impressions presented speeches by one hundred and twenty executives to a panel of one thousand listeners and ten experts and asked them for feedback. Its findings confirmed those of Mehrabian, though the numbers were slightly different than his formula predicted: The way the executives spoke accounted for 23 percent of how listeners evaluated the speakers. The actual content of their speeches accounted for only 11 percent. If this was right, then all of the ideas and experiences Sarah described in her job interviews were less than half as important as how she sounded when she spoke about them.

Sarah sent Quantified Impressions a video of herself speaking.

Analysts at Quantified Impressions ran the clip through software and technology that measured her tone and delivery against those of her peers.

It turned out that Sarah was an up-talker. You've heard this manner of speaking? Everything sounds like a question? Even when it's not? Every sentence ends on a note higher than the rest? It's common among the millennial generation. It's meant to keep a listener engaged, a way of saying *You know?* at the end of every sentence without actually speaking those words. But it frequently rubs older folks the wrong way or leads them to presume someone is younger or less experienced than he or she is. Sarah also learned that she spoke at a higher pitch when she was nervous. On her interview video, her voice frequency was measured at 263 hertz; the average female's is 216 hertz. She also rushed her words at the beginning of sentences, speaking at 220 words minute; the average among her peers was 170 words.

Quantified Impressions concluded that these three problems made an academic whiz with stellar recommendations, real-world experience, and incredible drive sound "young, inexperienced, insecure, and disingenuous." She was broadcasting a nonverbal message that directly contradicted the one she'd cultivated throughout her academic and professional life.

Sarah met with Briar Goldberg, a vocal coach who works with Quantified Impressions. Goldberg taught Sarah to identify situations that made her so nervous that her vocal pitch became elevated. To help with Sarah's up-talking, Goldberg assigned her rigorous enunciation exercises. Sarah was no stranger to homework. By the time her next round of interviews arrived, she had broken her bad speech patterns. She landed an internship with one of her top choices, a company located in the technology hotbed of San Francisco.

Effective nonverbal communication is also a valuable sales tactic. Superstars of motivational speaking — Tony Robbins, Brian Tracy, and Tom Hopkins, for example — have made millions talking to crowds. But for the past twenty years, when those guys need to sell tickets to their cash-cow events, they call on the "Phone Sales Guru," Gary Coleshill, who refers to himself on his website as the "undisputed king of cold calls." By 2013, Gary estimated, he'd made almost a million phone calls trying to sell people things — tickets to motivational seminars and more. His primary tool is his voice. He has dozens of phrases and conversational tactics he uses to keep people on the line and to make them interested in what he's selling. But at the beginning of every call, there's nothing more important than the pitch of his pitch.

Coming on too excitable screams *salesperson,* Gary says, so he always starts with a relatively casual tone. He's learned what vocal register works for all kinds of statements. For example, when you're issuing a command, he says, a deep voice works best. Questions get better responses when asked at a higher register. He's also very aware of the part of the country he's calling. In the morning, talking to New York, "I've got to get hard and fast and almost abrupt," he says. In the afternoon, he might call Charlotte, North Carolina. "I have to take a deep breath and slow it right down." The whole trick, he says, "is to make them feel like you're one of them. Make them comfortable like they're talking to one of their neighbors." Thing is, Gary's from England and speaks with a British accent. He sees his accent as an asset — the last thing Americans hear in a British accent is a salesperson. "Even though it's harder to understand me, they put more effort into it," Gary says. "They're a little more patient than they would be." That leeway is part of Gary's formula for success. It gives him just a little more time to convince his subjects that there's a kernel of truth in what he's saying about what he's selling. "You have to hook them fairly quickly," he says. "That's the essence of cold calls for me."

You can put Gary's tactics into practice when you deliver a talk or tell a story. If you are speaking quickly and rhythmically about a topic, try slowing down your tempo and pacing and becoming more emphatic just at the moment you are hitting your key point. This shift will surprise your listeners and get them to pay more attention. When Gary encounters someone who says he or she can't afford to buy whatever it is, he slows down and lowers his voice. "That's . . . precisely . . . why . . . this is the right time to buy," he might say. People are shocked. What? How so? "Prices will never . . . be . . . this . . . low . . . again."

Allison Dufty makes her living triggering emotions as a radio editor and voice artist. If you've visited the fourth floor of the Museum of Natural History in Manhattan recently, you've probably heard Allison talking about fossils and dinosaurs — her voice accompanies visuals on touchscreens there. She's also the voice on the audio tour at the Art Institute in Chicago and at the J. P. Getty Museum in Los Angeles, among others. She's the narrator on a popular boat tour around San Francisco Bay, and she's on the tram tour around Angel Island there. Once she did a series of recordings for a call-in health-advice line offered by health-care giant Kaiser Permanente. She forgot about the job, then years later she called the health line when she had a stubborn case of bronchitis. She ended up getting automated health advice from herself.

In her line of work, Allison spends a lot of time thinking about the way her voice sounds. The first time she recorded for a GPS navigation device, she says, she blew it. She recorded the messages in a kind of chipper voice — not gooey, but upbeat. Six months later, her client came back and told her the company was getting complaints from users. "Everybody said, 'She sounds too nice,'" Dufty recalls. Users wanted a more matter-of-fact voice. They didn't really want to hear her personality. As Dufty puts it, "Essentially, they wanted me to sound mean."

Allison's work requires her to pay attention to the basic building blocks of spoken language. These days, she often finds herself recording huge sets of snippets and sound bites in seemingly endless varieties. The words don't change from take to take, but the way they're spoken does. She speaks them to form both the beginnings and ends of sentences. Phrases end hanging on high notes in anticipation of other phrases latching on. Others resolve with low notes, indicating the ends of statements. With different inflections, single words can indicate either declarations or questions. Out of context, her recordings can sound like gibberish. But they're assembled to form answers like the ones you hear from Siri or Google Now (neither of those are Dufty's voice, to be clear). Pieces of those answers might actually have been recorded years apart, Allison says. Sound artists refer to this stringing together of sounds, phrases, and words as concatenation. "If you do it right, it sounds like talking," Dufty says. "If you do it wrong, it sounds like a ransom note."

We all need to be persuasive and communicate clearly, whether as CEOs, plumbers, friends, family members, construction workers, or, most certainly, teachers. It affects one's level of success in personal and professional relationships and can play a key role in advancement and job satisfaction. *How* people say things can be as important as *what* they say. When you are looking to persuade or prove a point, try deliberately slooooo-wing dowwwwwn at a key moment when you are communicating something important. Use the power of a brief silence, then change up your pace at the perfect moment to have your point be heard. Establish a cadence, then break it as soon as you arrive at your point. Talk in a lower register to signal authority, and raise the pitch of your voice to really engage someone with your question, just like Gary does.

Feeling wiped out after that big presentation? Researchers are constantly finding new wellness benefits in sound and music. There's no question about it: sound heals. We all instinctively know that a soothing familiar voice, song, or just a calm background or natural ambience can dramatically reduce stress, help manage pain, and increase a sense of well-being. For thousands of years, monks have used chimes, gongs, and other instruments with complex harmonic structures in rituals that connect them to the essential resonant frequencies that help define their experience as human beings, but much simpler approaches can be effective as well. Listening to bird sounds, even recorded ones, can lower your heart rate and calm you in just a few seconds. Just closing your eyes and focusing on the sound of your own active breaths can do the same. The sound of light laughter (especially your own) can cut tension or stress in a matter of moments.

Music therapy has been proven effective in some circumstances; for example, to reach or communicate with patients with psychological and brain issues like autism and Alzheimer's disease. Sometimes stroke patients or others recovering from brain injury can sing before they can talk, and for these people, singing is used in a treatment called melodic intonation therapy. (After being shot in an assassination attempt in January 2011, Arizona congresswoman Gabrielle Giffords recovered her ability to speak with the help of melodic intonation therapy.) In a November 1, 2013, study of sixteen stroke victims suffering from unilateral neglect (inattention to one side of the body, post-stroke), Taiwanese researchers found that playing these patients classical music appeared to improve visual attention more than playing white noise or nothing.

Even people who haven't suffered strokes can benefit from music therapy. Ever feel like your brain's fried after a day of going to back-to-back meetings or corralling kids? The research suggests your overall cognition improves with a musical break. Can you steal twenty minutes in an isolated corner or room with a pair of headphones? A sonic refresh might do wonders for your visual attention in the challenges you face for the rest of the day.

You might be a fan of Nike's Run app or others that match music with pace. But psychological studies have shown music can help you get more out of exercise. In a 2010 study, researchers from England played six popular music tracks to twelve healthy male students who cycled at their own chosen rates. Then the researchers sped up and slowed down the tempos of the music. They found that these test subjects not only chose to work harder on their bikes when the tempo increased but also "enjoyed the music more when it was played at a faster tempo." Music is good for exercise, which is good for you, and exercise can make you enjoy music more! So take a moment to create a workout playlist or make sure the service you're using to score your exercise knows to speed up the tempo when you need it most.

Costas Karageorghis of Brunel University in London has studied the psychological effects of music for twenty years. In a late 2011 review of research from the previous forty years, he wrote that music can make people less aware of fatigue and elevate mood, endurance, and metabolic efficiency. Music can even make people feel like they're not working as hard to achieve the same results they do without music. "In this sense, music can be thought of as a type of legal performance-enhancing drug," he wrote.

Years earlier, in fact, U.S.A. Track and Field, the national governing body for track and field, long-distance running, and race walking, decided to prohibit athletes from using iPods and music players while they competed "to ensure safety and to prevent runners from having a competitive edge." Runners protested, and a version of the rule now applies only to competitors who receive awards or money.

So you've got the go-ahead to get high on music for your next exercise, but I'm here to be your sonic pusher. Dope up the moment you wake up. Load up on sound for your daily walking commute from the parking lot, subway, or bus stop to the office. Give in to the urge to boost your productivity throughout the day, no matter what you're doing, with sound and music. Take a hit to relax before bed — beats sleeping pills! The only side effect is changing your perspective on what's possible when you harness the power of sound.

9

Hearing Around Corners

THERE ARE MOMENTS in the evolution of communication when very smart people transform the media all of us use every day and catapult us forward. A telegram becomes a telephone. The telephone becomes the gateway to the World Wide Web. The web becomes the basis of our connection to knowledge. The sharing of that knowledge online becomes the fabric of our connection to one another — a social web that inspires a radical shift in our notions of privacy, business, and storytelling. I like to think of technological and intellectual transition in terms of ideas that my grandmother never could have imagined in her lifetime, even though she lived to see many of them — or that I can't imagine now but that will be an inextricable part of my grandchildren's lives.

For early adopters who managed to navigate the privacy pitfalls, many of these shifts in thinking or technological advances have created opportunities for great wealth. But they've also fostered social change. Facebook helped mint plenty of young billionaires who leveraged all of the data we now share, but it also continues to serve as a powerful engine for social movements and helps more efficiently rally change agents into action. Sound, I believe, is the next frontier in business, storytelling, and movements. It's an untapped layer of opportunity on already robust modern communication networks (both digital and analog). Of course, people have used sound to advertise and market stuff for decades. But sound hasn't been harnessed at scale — as a tool for human connection. The people who realize this will benefit tremendously. We all will.

That's what this has all been about: recognizing and harnessing sound for your business, your cause, or your personal life. When you understand, as you now do, how sound works in your brain — when it's coming down your block offering ice cream with a tinkly tune or on your sizzling platter at dinnertime — there's no denying the power of sound to instantly get your attention and make you feel real emotions, recall memories, and take action. In fact, you now have a name for that sonic spark: the boom moment. You'll recognize boom moments now whenever the right sound at the right time sets off a barrage of input from other senses. You now understand how sound creates an important part of you daily landscape and how even small, functional, or personal sounds — from shoes, gunfire, and the crunch of food to sneezes, laughs, and the sound of people's voices — tell rich stories.

So dependable is the effect of sound that it can be harnessed with a set of principles you now know. You can tell the important difference between a jingle, which might get hammered into your long-term memory, and a full-fledged anthem, which captures a total story in the language of sound — and does so far more efficiently than most marketers with billion-dollar budgets realize. You have the advantage of seeing how the smartest brands and artists put the focus on the experience that sound and music create, not just on the sound or music itself. AT&T did it. So did NBC with the help of John Williams (and maybe even me and my team) for the Super Bowl. Martin Luther King Jr. harnessed sound to make history and showed generations of people a secret to emotionalizing and personalizing a social movement. Sound can even help a brand such as Univision tie into those types of social movements.

But you don't have to run a major corporation or lead a movement to transform your daily life with sound, from the moment you wake up to your daily commute to that big presentation to the restaurant you go to for dinner. Now, finally, you can start to see where some of the sounds I've already showed you could go in the near future. Businesses, movements, causes, and individuals could be doing some of these things with sound *tomorrow* if they wanted to — most of this technology already exists. When they do, they'll be able to create a better-sounding, better-working, more meaningful world. There's so much up for grabs in so many aspects of our lives.

SOUND AND SPACE

Sound is an essential part of any physical space. The sonic landscape gives a place purpose. Sounds in spaces can influence your velocity — how fast you walk, eat, or exercise (which is why you don't hear a lot of Perry Como in gyms). Sound tells instantly when a place is meant to be peaceful; sounds in spaces can create the perception of silence. The mere suggestion of sound can change the way you feel about a setting. Think about the difference you perceive at a restaurant when you're offered "a table in the corner" or "a quiet table in the corner."

We should be creating more public spaces as proofs of concept of the power of sound. Soundscaping should play just as much of a role as landscaping. A tree, after all, isn't just an effective way to make grounds look prettier or deliver more oxygen. It attracts birds. And birds make sounds that we associate with peace and calm. Leaves rustle in the wind. Water features aren't just for looks either. They're home to creatures whose sounds tell stories about times of day, climates, and seasons. Running water is in that same Zen-feeling palette of peaceful sound. Many of those same natural sounds can be harnessed indoors or just outside of our windows to the same ends. In urban environments, the concept of a park could be altogether reimagined: we could build sonic parks full of curated boom moments.

Businesses could do this all over office spaces. In an age where the open plan rules and fewer and fewer people can shut office doors for private moments, break rooms or private spaces could be organized by need states, moods, or feelings and serve as quick stops for anyone who wants to transform or reboot his or her day — they wouldn't always have to calm you down; they could rev you up or energize you for a big meeting or for exercise or help you wind down before heading home.

Or consider the power of sound to create virtual silence in the same setting: Some smart offices (such as the headquarters of financial research company Morningstar in Chicago) with open plans broadcast directional white noise from the ceiling down onto cubicle spaces, creating a noise-canceling virtual cone of silence that allows people within feet of each other to carry on simultaneous phone or in-person conversations without annoying each other.

Sound can also influence our actions in those spaces, including the way we buy. As retail spaces compete with the e-shopping experience, they have to become much smarter with sound to survive. They could give you a chance to open up your social graph so that when you come in to a store, a lobby, or a department, *your* soundtrack would play — on headphones or overhead, depending on the size of the store. You'd feel catered to, like this store was *your* store. Privacy would be an important hurdle. But consider all we already give up *without* the return of a more personal, pleasurable real-life experience. We let Facebook track our music-streaming habits on Spotify and other services so our friends can see what we listen to and then comment on it or like it. We share our every move in the form of location data with streaming radio services such as Pandora or Google Music. They're using that data to their advantage. And why not? We grant them permission. As we expect more tailored experiences, we're increasingly willing to share data about ourselves. And music is a portal into a powerful means of personal expression.

The implications of personalized sound for advertising and marketing are simultaneously scary and intriguing. Imagine you're in a crowded mall or shopping district. Blinking lights and animated signs beg for your visual attention. Bright hues insist you buy one to get one free. The occasional snippet of a song or technological beep cuts through the cacophony. Suddenly, your smartphone buzzes. You open it up and see a push notification from an app — call it the Sonic Boom app — and it's showing you which brands are using sound to tell you something. (As this book was being written, Shazam, an app that first debuted as a way to use your smartphone to identify and purchase music heard in the wild, added a feature that let your phone automatically listen constantly for songs and brands.) Maybe this imagined app offers a discount on the brands it hears. Or a map. Or directions. The instance of a brand sound is part of your daily soundtrack, which the app also keeps track of. An app like this could automatically begin to filter out the noise — the sounds that aren't telling you anything or guiding you in any way. Plenty of people would share some personal info for that considerably less noisy experience.

SONIC USER INTERFACES

There are all kinds of sonic opportunities left behind when it comes to the design of user interfaces. They're the ways we access technology. And the goal of the best user interfaces is often to make the technology itself disappear into an experience. "It just works" is how Steve Jobs described the user interfaces of Apple's devices.

People play with sonic triggers in user interfaces, but too often those sounds don't benefit anyone, don't mean anything, or don't tell a larger story. They're neat, at best. Take, for example, the way scientists and designers have sonified the volume of trading measured by the Dow Jones Industrial Average. But why don't Wall Street traders use these kinds of sonic tools to help convey an urgency when a price dips or spikes so they can act accordingly? Aren't traders analogous to the fighter pilots who use sounds to warn of incoming enemies in this type of situation? Chefs or home cooks managing multipart dinners could be using different intuitive notification sounds to tell them when various courses were done. Your calendar could tell you when you have a certain number of minutes until your next meeting with a helpful sound instead of the intrusive paragraph of text that pops up in the middle of your work after a clunky *bonk*. Sound could create the landscape in which we make these important decisions and lead to better, more considered outcomes.

On a lighter note, sound in user experiences could save us time and effort. Think about your last trip to your bank's ATM. If it didn't make a sound, how hard did you have to watch the screen to make sure the buttons you pressed actually registered? That same system could be part of a rich sonic strategy for a financial institution. What if more banks created better sonic identities? Like AT&T's anthem, a bank's score could capture all of the distinguishing characteristics of the brand: the people's bank; the world's most convenient bank; the world's most trusted bank, and so on. The anthem, which would be heard on TV and radio, online, and everywhere the bank brand appeared, could be distilled into smaller instances of sound that would show up wherever the bank enabled a transaction. At the ATM, each progressive note of your four- or five-digit PIN could be a note in the theme of the bank's anthem. Getting money would become a satisfying game: Your individual PIN would always be unique, and no one could hear the numbers you'd typed, but the first, second, third, and fourth numbers (whatever they were) would always play the notes of the familiar sonic bank logo as you made your secure transaction. Your brain would hanker to complete the little song because it craves familiar patterns, and you'd have heard this one all over the place. Plus, you'd know not only whether you'd actually pressed the button hard enough but, without looking or paying that much attention, whether you were on the first, second, third, fourth, or fifth number of the PIN. And when you'd finished playing the little song, you'd get cash. The feedback loop would continue every time you heard that sound — as part of an anthem or even during an otherwise inconvenient time spent on hold with customer service. You'd remember that feeling of getting what you needed, and the sound would reinforce a positive experience.

My company recently had the good fortune to work with Weather Channel, which owns and runs one of the most popular mobile apps, with over a hundred million downloads and counting at the time of this writing. We created alert sounds that let you know (if you opt in, of course) when rain — or something more threatening, like a hurricane or tornado — is on the way. And the signal becomes more insistent with the severity of the weather event, sounding almost like an alarm or warning that sonically suggests you hightail it out of wherever you are and run for cover. The point is, you *feel* the message of impending threatening weather before you even take your phone out to look at the details. The sound helps you instantly know what level of attention is required.

In an age where we're buying more and more online, digital stores should be more deliberate about their use of sound. The shopping website Gilt.com is pushing into this space with mobile alerts it sends to registered customers who opt in. They hear the energizing, urgent-sounding notification play the instant a flash sale starts. Each day, thousands of aficionados *love* the rush of excitement that's triggered when they hear that sound. What if online stores played a sound when you unlocked a special VIP experience or discounts on luxury items? As long as you were already in a receptive frame of mind, the right sound could subtly nudge you along in the shopping experience.

DATA SONIFICATION

Laboratories, hospitals, social science centers, and universities could glean insights from the sonification of data. Hospitals already suffer from what NPR identified in a January 2014 story as "alarm fatigue" — too many alarms are sounding at any given moment in a

hospital to create the right sense of urgency, much less offer a meaningful story about precisely what's happening with a patient. "The
three-burst is a crisis alarm," systems engineer James Piepenbrink of
Boston Medical Center told NPR—it means a life-and-death situation. But "two tones is a warning," he said, which could signify a
whole range of potentially fatal or relatively harmless things. The
hospital's solution was to silence more alarms. But why waste all of
that valuable data just because it's attached to dumb sound? Even
simple rising or falling tones or varying velocities of ticks or clicks
could tell a richer story.

We've only recently become enthralled with the notion of presenting data visualizations as a way to reveal hidden insights. Infograph-
ics, though mostly ill-considered, are all the rage in design circles,
and they're among the most viral content on the web. The ones that
stick are beautiful from the get-go. They play with color or geometry
in a way that appeals to the eye and to the patterns and broken patterns the brain craves. These visuals draw us into stories that data
tells in ways we'd overlook if they were just numbers or words on a
page. Sound can have the same effect, but it works even quicker. An
interesting sound that comes loaded with a set of feelings or indica-
tions—a rising or falling tone; a quickening or slowing pace (to put
things very basically)—could garner the same kind of attention that
engaging visuals do. But they'd work even if we were looking in the
other direction. Presenting data as sound is the next logical step in
the ever-accelerating information age.

More could be done in the near future with sound that is, literally,
the sound of you: your DNA, sonified. In 1979, Pulitzer Prize–winning author Douglas R. Hofstadter started talking about DNA/RNA
sequencing in terms of a tape recorder. And years later, scientists
and composers began sonifying these sequences. More recently, the
sonification of human DNA and RNA has been proposed as a way to
compare and analyze it.

Right now, the majority of our identifying documents are either based on or heavily reliant on visual elements. More and more, certain access codes are built on sounds — often the sounds of people's voices. But even that presents problems (what if someone has a cold, and what happens with children whose voices change as they age?). Admitting there would be potential for fraud and misuses, what if the primary means of universal identification was one's DNA sequence, and that DNA sequence was represented by a kind of song? It could come up everywhere — at the bank, in your car, on your mobile device, on your home-security system.

THE FUTURE OF PERSONAL SOUND

James Bond and *Saturday Night Fever*–era John Travolta aren't the only ones who get to have personal soundtracks. Modern music-streaming services and devices let you dial up virtually any song in the world wherever you go. But there's always music playing in your head too — even in the form of earworms that you might not think you like but that actually help you get through mundane tasks every day.

But an even deeper sound of *you* could be at your fingertips, no matter where you are. You would be able to pull your personal sonic triggers several times a day to quickly convey needs or preferences and set your mood. It's already part of the sonic identity you constantly express — from the sound of your voice to the sound of your shoes — so why not refine and amplify it?

Your personal soundtrack could follow you everywhere. The seams between music platforms and spaces where your music plays will eventually disappear; a playlist need not stop just because you go from your office to your car to your garage to your living room.

What if you thought of a mobile phone as a speaker in your pocket or purse? Except that it's a speaker with software that happens to know where you are (if you let it), whether you are traveling or standing still, plus the time and date and some personal information, like your birthday and the birthdays of your friends and family. Check your Facebook or LinkedIn settings — you're probably already sharing this stuff on those social networks. Maybe that network has collected intel on the kinds of music you stream. If you chose to opt in — and you probably did if you use virtually any streaming music or social media app these days (read your terms of service!), an app could automatically create a soundscape that would shift and morph with your circumstances. You might take issue with the level of privacy you were giving up, but the actual experience of such an app might feel more magical than invasive. It would help score your life in a way that instantly felt meaningful. You'd be offered ocean waves or white noise to help you sleep, a killer beach playlist or a gentle Sunday wake-up mix — some services, such as Beats Music, Songza, SoundTracking, and others are flirting with the concept by asking you to manually input your mood, activities, or specific locations in natural language.

Someday soon you could passively collect a travelogue of sound from your vacation to help you remember how you felt at each stop, or perhaps you could capture an audio bank of voices of loved ones and friends and then call them up to instantly make you feel the warmth of their company. Cueing up the feelings you associate with mom, dad, brother, sister, son, daughter, or boyfriend need involve only the tap of a button.

What if, on a human level, we shared more sets of sounds to signify all sorts of important messages and feelings? Sounds or songs could represent information you chose to share — your address, phone number, e-mail, and more. All of us could have a universal sonic language that crossed cultural and language barriers. You can pick out the sign for a public restroom pretty much all over the world; what if there were universal sonic symbols for "walk," "don't walk," "speed up," "slow down," "danger ahead," "police," "fire," "emergency exit," "stop," and "go"? World travel would get a lot easier and less stressful. We'd just communicate better: Instead of trying to figure out the word for "toilet" in English, Japanese, or Farsi, you'd simply whistle a tune to a stranger. Call it a kind of sonic Esperanto.

Your sound and music could show up in any environment where it might make a difference. Unless silence was needed. And then the right sound would help create that perception of silence.

Imagine the low-level white-noise barrier used by businesses applied in your personal life. It could create a buffer between the TV room and the dining room, so noise from the kids' cartoons wouldn't compete with dinner conversation. The noise of freeway traffic whizzing by a neighborhood could magically disappear when counteracted by anti-road noise, sound on a frequency that cancels the frequency of road noise. The noise-canceling sound need only be piped through speakers in a small buffer zone.

For all of the benefits to business that sound offers, there's no greater impact than the pleasure or energy sound can bring to your personal life. It's nothing short of a tool for happiness, one that's available to virtually anyone with a functioning sense of hearing — and a few of the tools of understanding in this book.

At the beginning of this book I promised you'd never hear the world the same way again. I hope you've already started to notice things you never noticed before — sounds you never knew were there, the emotions or memories they trigger, the way they transform the mood in a room or the direction of a conversation.

This book is a field guide to making a better-sounding future now. Now that you've learned how to recognize the sound all around you, you'll hear your own set of opportunities. I hope you'll continue to experiment with creating your own boom moments. Even after my many years of exploring and innovating in sound and music and collaborating with so many brilliant creative minds, I'm still learning new things every day about the powerful and pervasive influence of sound.

Michelangelo famously said, "Every block of stone has a statue inside it and it is the task of the sculptor to discover it." The same concept applies to sound in our lives. We have to chisel through layers of noise around our ideal sonic experiences to find the opportunities for boom moments.

Start now. When you put this book down, close your eyes and think about all the layers of sound you hear. Those sounds have the power to make you feel something. They have stories they're trying to tell you. There's a boom moment in there somewhere.

Acknowledgments

Joel first and foremost expresses heartfelt gratitude and thanks to Tyler Gray for his boundless energy and enthusiasm, the endless fountain of ideas in his brain, and his finely tuned listening and storytelling skills. Without his heroic efforts and passion to help bring these ideas to the page, this book would have been so much less than it is.

To my wife, Tracy, and amazing children, Josh and Emily, as my greatest sources of joy and inspiration, and for their patience and understanding for all the times their husband/father was lost in thought in the sound, or in the pages. I also want to acknowledge, with much love, my mother, Nancy, who enriched our house and triggered my curiosity with the infectious sounds of Stevie Wonder and Laura Nyro; and my father, Stephen, who lovingly impressed upon me the power of inquiry and perseverance. My siblings, Ken and Wendy, each musically talented in his and her own right, are always there for frequent advice and steadfast support and belief.

I also acknowledge the dream team that is my entire staff at Man Made Music, under the leadership and creativity of Kim Paster, Allison Meiresonne, Lauren McGuire, and Dan Venne. Their collective sense of curiosity, creativity, and drive inspires me and helps me learn something new every day. Gratitude also to Julia Padawer, who helped bring some important strategic thinking to the early sonic work, and to Morgan Inman, who years ago came upon the rudiments of the Mustang story included herein.

Special thanks are due to my frequent and long-time musical partners-in-crime Phil Hernandez and Chris Maxwell, Dennis Wall, and Jim Hynes, for their friendship, stewardship, and unerring sense of heart and musicality. You keep me honest. My truth barometers and mentors also include Rose Hirschorn, for providing astounding insight into the elemental instrument of voice, and Richie Becker, for his mind-expanding explorations into the relentless pursuit of great sound that moves people.

To all the creative souls, the incredible music creators, artists, bands, consummate musicians, directors, show runners, writers, executives, designers, producers, and editors with whom I have had the immense pleasure and honor to work—I've learned so much from you about art and storytelling that I've attempted to apply here in these pages. There is no way to thank you all, but I wish I could. Your work brings the world forward, and fills my creative life to the brim.

To the unflappable Lynn Johnston, who was an elemental force in bringing the book to life and who carried me on her back through the trials and tribulations of publication, and to Pam Workman, (along with the amazing team at Workman Group), who connected me with Tyler Gray and who has for years made a big impact in the marketing of my company and now the book. Grateful nods are also due to Jay Cooper for his role in helping me grow my career, John Newsholme for expert financial and business management, and especially Victor Metsch for his masterly strategies that helped keep the ship on an even keel in stormy seas. Special thanks also to Thom MacFarlane for his insightful view on the content from an education perspective, which helped elevate the book, and to my longtime adviser and coach David Dowd for his many years of boundless support and inquiry.

A very big thanks from both of us to all our friends at Houghton Mifflin Harcourt who truly shared the vision from the beginning, under the strong leadership of Bruce Nichols, including Lori Glazer,

Katrina Kruse, Stephanie Kim, Carla Gray, and Naomi Gibbs. A special deep debt of gratitude is owed to our tireless editor, Courtney Young, who pushed us far outside our comfort zone and went way beyond the call of duty in bringing the gift of clarity, focus, and rigor to the prose. Her efforts shine throughout these pages.

We would like to express thanks to Taylor Beck and Kristen Lueck for researching the neuroscience and psychology, much of which went into the One-Second Brain Science infographic on page 25. Special thanks to Aaron Harvey, founder of digital design firm Ready Set Rocket in New York, and designer Anastasia Kuznetsova for turning that research into something beautiful.

Tyler would like to thank his mother, Mary Lynne Gray, and his late father, Michael Gray, for encouraging music in his life. And he'd like to thank his wife, Sia Michel, and their children, Cash, Ida, and Beau Gray, for their patience, support, and genuine curiosity throughout the process of writing this book. He would also like to thank Noah Robischon and Robert Safian for allowing him flexibility and supporting his work on this book while he was at Fast Company.

Notes

Introduction

Page

xv   The Shining: Watch the video here: https://www.youtube.com/watch?v=KmkV
     WuP_sOo.

1. Night-Vision Goggles for Your Ears

2   *"Twenty-nine years old and hearing myself"*: Watch Sarah's original video here:
     http://www.youtube.com/watch?v=LsOo3jzkhYA. A video of her appearance on
     Ellen DeGeneres's talk show can be found here: http://www.youtube.com/watch
     ?v=fp4usWroDew.

5   *"Public toilets are insanely loud"*: Some of these quotes are from Sarah's memoir,
     but she was also interviewed by the authors via e-mail in March 2013.

6   *ability to detect sound vibrations:* This information comes from *The Universal
     Sense: How Hearing Shapes the Mind,* by Brown University neuroscientist Seth S.
     Horowitz (New York: Bloomsbury USA, 2012).

7   *24 percent:* See Mike J. Dixon et al., "Losses Disguised as Wins in Modern Multi-
     Line Video Slot Machines," *Addiction* 105 (October 2010): 1819–24. The research-
     ers used skin conductance responses to measure the effects of music and sound on
     excitement. Music and sound caused a spike in the estimation of winnings of from
     15 to 24 percent.

13   *Reekes says:* I talked to Jim Reekes a lot in person and via e-mail, but here's
     one of the first times he told a version of this story: http://www.youtube.com/
     watch?v=QkTwNerh1G8#at=1162. The video was uploaded February 23, 2010, by
     One More Thing, an Apple community in the Netherlands, Belgium, and Luxem-
     bourg.

16   *Ellen DeGeneres's show:* Watch Sarah Churman on Ellen DeGeneres here (get a tissue first): http://ellen.warnerbros.com/videos/index.php?mediaKey=1_ccmh 4z8w.

2. The Boom Moment

19   *Juan Antonio "Sonny" Falcon:* I interviewed Sonny Falcon by phone. He continues to judge fajita cook-offs around Texas.

21   *the "fajita effect":* Chris Brogan of publishing and media company Human Business Works might be the one to have coined the term; it appeared in an April 11, 2011, blog post titled "Smell the Sizzling Fajita." Read it here: http://www.chrisbrogan .com/fajitas/.

23   *orienting response:* See the book *Sweet Anticipation: Music and the Psychology of Expectation*, by David Huron (Cambridge, MA: MIT Press, 2006).

hippocampus and amygdala: Again, Seth Horowitz's *The Universal Sense* was particularly informative on this issue — particularly chapter 5, "What Lies Below: Time, Tension, and Emotion." He takes on the hard-to-believe idea that only sound, touch, and being thrown off balance can cause a true startle. Key to this argument is understanding precisely what happens in a genuine startle reflex, in which a very specific set of synapses fire in a fraction of a second.

24   *Janata says:* Janata isn't just riffing on a few fMRI studies. He's one of the foremost experts in the field. He has written extensively on the subjects of music, emotion, and memory. Among the publications he co-authored that informed this section are "Characterization of Music-Evoked Autobiographical Memories," a 2007 study with S. T. Tomic and S. K. Rakowski published in the journal *Memory;* "The Neural Architecture of Music-Evoked Autobiographical Memories," published in 2009 in *Cerebral Cortex;* and "Music-Evoked Nostalgia: Affect, Memory, and Personality," a 2010 article with F. S. Barrett, J. K. Grimm, R. W. Robins, and T. Wildschut published in *Emotion.* Links to them all, plus several other news stories about Janata and his work, can be found here: http://atonal.ucdavis.edu/projects/ memory_emotion/.

25   *We react faster to sound:* See B. J. Kemp, "Reaction Time of Young and Elderly Subjects in Relation to Perceptual Deprivation and Signal-On Versus Signal-Off Condition," *Developmental Psychology* 8 (1973): 268–72.

Elite sprinter's leg muscles: See M.T.G Pain and A. Hibbs, "Sprint Starts and the Minimum Auditory Reaction Time," *Journal of Sport Sciences* 25 (1), (2007): 79–86. See also Leila Nuri, Azadeh Shadmehr, Nastaran Ghotbi, and Behrouz Attarbashi Moghadam, "Reaction Time and Anticipatory Skill of Athletes in Open and Closed Skill-Dominated Sport," *European Journal of Sport Sciences* 13 (5), (2013): 431–36.

Simple reaction to sight: See *Stanford Encyclopedia of Philosophy,* "Some Relevant

Empirical Findings (Psychology, Psychophysics, Neuroscience)": http://plato
.stanford.edu/entries/consciousness-temporal/empirical-findings.html.

Simple reaction to touch: See Ingrid M.L.C. Vogels, "Detection of Temporal Delays
in Visual-Haptic Interfaces," *Human Factors* (2004): 46, 118–134.

Beginning of being able to recognize music: See Alexander Refsum Jensenius, "How
Do We Recognize a Song in One Second? The Importance of Salience and Sound
in Music Perception" (Thesis, University of Oslo, 2002).

27 *"Burt's choice of"*: This is from a piece by Daniel Neely for a forthcoming book:
Sumanth Gopinath and Jason Stanvek, eds., *The Oxford Handbook of Mobile Music Studies,* volume 2 (New York: Oxford University Press), 146–71.

28 *C. Phillip Beaman:* Particularly informative was "Earworms ('Stuck Song Syn-
drome'): Towards a Natural History of Intrusive Thoughts" by C. P. Beaman and
T. I. Williams, published in 2010 in the *British Journal of Psychology.* A pdf is available here: http://centaur.reading.ac.uk/5755/1/earworms_write-upBJP.pdf.

28 *Andrea Halpern:* I spoke at length with Andrea Halpern via Skype video call about
this topic. That conversation helped inform this and several other sections about
earworms and recall and prediction involving music.

Ira Hyman: I spoke extensively with Ira Hyman via phone. His January 19, 2013,
article, part of his Mental Mishaps series titled "Goldilocks and Controlling Intrusive Thoughts," can be found here: http://www.psychologytoday.com/blog/mental
-mishaps/201301/goldilocks-and-controlling-intrusive-thoughts. More on Hyman
himself can be found here: http://www.psychologytoday.com/experts/ira-hyman.

29 *Baruch College:* The findings of the January 2006 Baruch College Special Report
#9, "Neighborhood Noise and Its Consequences," referenced in a May 20, 2007,
New York Times article titled "As Ice Cream Trucks Tune Up with Songs That Mad-
den or Gladden," can be found here: http://www.etownpanel.com/documents/
Special%20Report_9_%20January2006.pdf.

By 1949: This historic information comes via Daniel Neely from Eleanor Harris's
August 20, 1949, article "The Pied Pipers of Ice Cream" in the *Saturday Evening
Post.*

3. Sonic Landscapes

42 *"systems with more aggressive"*: From the 2009 National Highway Traffic Safety
Administration study entitled "Effectiveness and Acceptance of Enhanced Seat
Belt Reminder Systems: Characteristics of Optimal Reminder Systems." Find it
here: http://www.nhtsa.gov/DOT/NHTSA/NRD/Multimedia/PDFs/Human%20
Factors/Reducing%20Unsafe%20behaviors/811097.pdf.

43 *Ghost Army:* Rick Beyer made an award-winning documentary called *Ghost Army*
about the unit that used sound and other sensory tactics to deceive enemies. It
aired on PBS in 2013. Find it here: http://www.ghostarmy.org/.

44   *Recette:* Jesse Schenker told this story in an April 23, 2012, article in the *New York Times.*

45   *the effect of noise on taste:* Unilever's announcement about the findings is here: http://www.unileverusa.com/innovation/researchdiscoveries/sound/. It's based on the full white paper titled "Effect of Background Noise on Food Perception," by A. T. Woods et al.

46   *Spence and Blumenthal:* Barb Stuckey wrote about Charles Spence's work with Heston Blumenthal in a March 11, 2012, story called "The Taste of Sound" in the online publication *Salon.* Find it here: http://www.salon.com/2012/03/11/the_taste_of_sound/.

47   *Ronald E. Millman:* Sound is particularly important to Ronald E. Millman, in part because he is blind.

49   *flagship innovation store:* Find images of AT&T's Chicago Innovation store here: http://www.callison.com/index.php/at-t-michigan-avenue. The story from *Crain's Chicago Business* is here: http://www.chicagobusiness.com/article/20120829/BLOGS01/120829747/nope-not-an-apple-store-at-ts-michigan-avenue-flagship-aims-high#.

54   *number-one restrooms:* The full citation is here: http://www.bestrestroom.com/us/Hall_of_Fame/2007/junglejims.asp.

4. The Principles of Sonic Branding

59   *study by Japanese researchers:* "Sad Music Induces Pleasant Emotion," by Ai Kawakami, Kiyoshi Furukawa, Kentaro Katahira, and Kazuo Okanoya, was published on June 13, 2013, in the journal *Frontiers in Psychology.* Find it here: http://www.ncbi.nlm.nih.gov/pmc/articles/PMC3682130/.
    "the Good Enough Revolution": Robert Capps's August 24, 2009, story in *Wired* magazine, "The Good Enough Revolution: When Cheap and Simple Is Just Fine," can be found here: http://archive.wired.com/gadgets/miscellaneous/magazine/17-09/ff_goodenough?currentPage=all.

61   *"I think our human ears":* Joseph Plambeck's March 9, 2010, story in the *New York Times* called "In Mobile Age, Sound Quality Steps Back" can be found here: http://www.nytimes.com/2010/05/10/business/media/10audio.html.
    "Bass has always been": Jesse Dorris's September 11, 2013, story in Slate.com called "The Beats with a Billion Eyes" can be found here: http://www.slate.com/articles/technology/technology/2013/09/beats_by_dre_market_share_how_the_headphones_company_conquered_the_market.html.

64   *According to the NPD Group:* Michael De La Merced reported on Beats' financials in a September 27, 2013, *New York Times* story called "Beats Gets Infusion of Capital from Carlyle Group"; see http://dealbook.nytimes.com/2013/09/27/beats-secures-investment-from-carlyle-and-buys-out-htc/.

71   *NHTSA:* The National Highway Traffic Safety Administration's September 2009

study (DOT HS 811 204) "Incidence of Pedestrian and Bicyclist Crashes by Hybrid Electric Passenger Vehicles" can be found here: http://www-nrd.nhtsa.dot .gov/Pubs/811204.PDF.

80   *$2.1 billion business:* Sales statistics for ringtones were reported by Fierce Mobile IT. They can be found here: http://www.fiercemobileit.com/press-releases/ gartner-says-worldwide-online-music-revenue-end-user-spending-pace- total-63.

5. Rethinking Possibilities

86   *5,000 percent:* AT&T press release on June 28, 2010, titled "AT&T Expands Wireless Capacity for 3G Mobile Broadband Network in Manhattan and Throughout NYC."

87   *Wanda T. Wallace:* Wallace's articles "Memory for Melodies: The Effect of Learning Music and Text Together" and "Jingles in Advertisements: Can They Improve Recall" were helpful to this section. Those articles and others can be found here: http://www.acrwebsite.org/search/view-conference-proceedings.aspx?Id=7167.

88   *The Art of Digital Music:* More on the book here: http://www.artofdigitalmusic .com/.

91   *Marissa Mayer:* The story of Marissa Mayer's forty-one shades of blue was first reported in the *New York Times* on February 28, 2009. Find it here: http://www .nytimes.com/2009/03/01/business/01marissa.html?pagewanted=3.
      internal leaked documents: Find the leaked documents of the Arnell Group's Pepsi logo rationale here: http://adage.com/images/random/0209/pepsi-arnello 21109.pdf.

6. Amplifying Messages

104   *Hispanic and Latino population:* Statistics on the composition of the Hispanic population can be found via the Pew Research Hispanic Trends Project here: http://www.pewhispanic.org/2013/06/19/hispanics-of-mexican-origin-in-the- united-states-2011/.
      distinguishing values: Several key Pew studies referenced here can be found at the following websites:

      http://www.pewhispanic.org/2012/04/04/iii-the-american-experience/
      http://www.pewhispanic.org/2013/07/23/a-growing-share-of-latinos-get-their- news-in-english/
      http://www.pewsocialtrends.org/2013/02/07/second-generation-americans/
      http://www.pewhispanic.org/2013/07/23/a-growing-share-of-latinos-get-their- news-in-english/.

109   *"Europe":* See a BBC News report on the decision about the Kosovo national anthem here: http://news.bbc.co.uk/2/hi/europe/7292240.stm.

112   *"I don't have any sense":* A transcript of the August 1, 2012, edition of *Talk of the Nation* can be found here: http://www.npr.org/templates/transcript/transcript.php?storyId=123599617.

      Freedom songs: The Spirituals Project at the University of Denver informed this section. Listen to the project's recordings of freedom songs here: http://www.spiritualsproject.org/sweetchariot/Freedom/civil.php.

7. Scoring the Experience

123   *Sukhbinder Kumar:* Kumar gave me a preview in an interview and via e-mail of a study he was working on at the time of this writing having to do with valence sounds in unexpected categories — sounds that don't fit on a spectrum of natural sounds that are annoying based on their frequency alone (he mentioned chewing sounds in particular). A key paper in helping put together this section is Sukhbinder Kumar et al., "Features versus Feelings: Dissociable Representations of the Acoustic Features and Valence of Aversive Sounds," *Journal of Neuroscience,* August 9, 2012. Find it here: http://www.jneurosci.org/content/32/41/14184.

125   *Hyman says:* Hyman referred to a November 1999 study by M. D. Schulkind, L. K. Hennis, and D. C. Rubin titled "Music, Emotion, and Autobiographical Memory: They're Playing Your Song." Find the abstract and download the link here: http://www.ncbi.nlm.nih.gov/pubmed/10586571.

136   *according to Nielsen:* Find those Nielsen numbers here: http://www.sbnation.com/2012-super-bowl/2012/2/6/2775693/super-bowl-viewership-record-madonna-halftime-show.

8. Creating Boom Moments Every Day

145   *"Is Noise Always Bad":* See Ravi Mehta, Rui (Juliet) Zhu, and Amar Cheema, "Is Noise Always Bad? Exploring the Effects of Ambient Noise on Creative Cognition," *Journal of Consumer Research* 39 (December 2012); http://www.jstor.org/stable/10.1086/665048.

146   *Phonics:* See L. C. Ehri et al., "Systematic Phonics Instruction Helps Students Learn to Read: Evidence from the National Reading Panel's Meta-Analysis," *Review of Educational Research* 71 (2001); excerpted in relevant reviews at http://www.cckm.ca/CLR/phonics.htm and http://www.dyslexie.lu/JDI_02_02_04.pdf.

      Lessons are picked up: A thorough review of contemporary research on phonics versus whole-language study can be found at the Heinemann website (a division of Houghton Mifflin Harcourt, the publisher of this book): https://www.heinemann.com/shared/onlineresources/08894/08894f2.html.

154   *melodic intonation therapy:* For a thorough explanation of melodic intonation therapy, see the July 2009 National Institutes of Health paper "Melodic Intonation

Therapy: Shared Insights on How It Is Done and Why It Might Help," by Andrea Norton et al.; http://www.ncbi.nlm.nih.gov/pmc/articles/PMC2780359/.

Taiwanese researchers found: "Listening to Classical Music Ameliorates Unilateral Neglect After Stroke," by Pei-Luen Tsai et al., was published in the *American Journal of Occupational Therapy* in the May/June 2013 issue. Find it here: http://ajot .aotapress.net/content/67/3/328.

exercise can make you enjoy music more: J. Waterhouse, P. Hudson, and B. Edwards, "Effects of Music Tempo upon Submaximal Cycling Performance," *Scandinavian Journal of Medicine and Science in Sports* (August 2010); http://www.ncbi .nlm.nih.gov/pubmed/19793214#.

9. Hearing Around Corners

163 *"alarm fatigue":* Richard Knox, "Silencing Many Hospital Alarms Leads to Better Health Care," NPR, January 27, 2014; http://www.npr.org/blogs/health/2014 /01/24/265702152/silencing-many-hospital-alarms-leads-to-better-health-care.

sonification of human DNA and RNA: Charlie Cullen and Eugene Coyle, "Rhythmic Parsing of Sonified DNA and RNA Sequences," Dublin Institute of Technology white paper, January 1, 2003; http://arrow.dit.ie/dmccon/25/.

Index